Giulio Donato Broccoli

The graph of a function

56 functions with solutions
+
120 exercises

Multi-lingual
English - Spanish – French – German – Italian

Editore Giulio D. Broccoli

Proprietà letteraria riservata

*Ogni riproduzione, con qualsiasi mezzo
(fotocopie, microfilm, microbiche, ..., ecc.)
totale o parziale è vietata.*

First Edition October 2009
Lulu.com
ISBN 978-1-4452-0827-5

New Edition 2010 - Italy
Editore Giulio D. Broccoli
ISBN 9788890530159

www.matematicus.com

To my parents
A mis padres
Ai miei genitori

INDEX

Dictionary   Page 7
1. English - Spanish   9
2. English - French   11
3. English - German   13
4. English - Italian   15

Graphs of functions   Page 17
1. Graphs of polynomial functions   19
Exercises
2. Graphs of fractional rational functions   31
Exercises
3. Graphs of irrational functions   57
Exercises
4. Graphs of trigonometric functions   77
Exercises
5. Graphs of exponential and logarithmic functions   107
Exercises
6. Graphs of functions with absolute value   131
Exercises
7. Graphs of other functions   143
Exercises

Appendix   169

Dictionary

- English – Spanish
- English – French
- English – German
- English – Italian

English → Spanish

The graph of a function	La gráfica de una función
56 functions with solutions + 120 exercises	56 funciónes con soluciones + 120 ejercicios
Graphs of polynomial functions	Gráfica de funciones polinómicas
Graphs of fractional rational functions	Gráfica de funciones racionales
Graphs of irrational functions	Gráfica de f funciones radicales
Graphs of trigonometric functions	Gráfica de funciones trigonométricas
Graphs of exponential and logarithmic functions	Gráfica de funciones esponenciales y logarítmicas
Graphs of absolute value functions	Gráfica de funciones en valor absoluto
Graphs of other functions	Gráfica de otras funciones
Graph the function	Representar gráficamente la función
Graph the functions	Representar gráficamente las funciones

DOMAIN — **CAMPO DE EXISTENCIA** (conjunto de definición))

SYMMETRIES AND PERIODICITIES — **SIMETRIAS Y PERIODICIDAD**

even function	función par
odd function	función impar
periodic function	función periódica
period	período
is an even function	es una función par
is an odd function	es una función impar
is a periodic function	es una función periódica

SIGN OF FUNCTION — **SIGNO DE LA FUNCIÓN**

INTERSECTIONS OF THE GRAPH WITH THE COORDINATE AXES — **CORTE CON LOS EJES**

LIMITS AND ASYMPTOTES — **LÍMITES Y ASÍNTOTAS**

vertical asymptote	asíntota vertical
horizontal asymptote	asíntota horizontal
oblique asymptote	asíntota oblícua
removable discontinuity	discontinuidad evitable
jump discontinuity	discontinuidad por salto (I especie)

THE FIRST ORDER DERIVATIVE — **PRIMERA DERIVADA**

Derivative	Derivada
Local minimum	mínimo local
Local maximum	máximo local
Local minimums and maximums	Mínimos y máximos locales
y is increasing	y es creciente
y is decreasing	y es decreciente

THE SECOND ORDER DERIVATIVE — **SECUNDA DERIVADA**

Inflection point	Punto de inflexión
Inflection points	Puntos de inflexión
y is convex	y es convexa o cóncava hacia arriba
y is concave	y es concava o cóncava hacia abajo
The graph of function is represented in Figure	La gráfica de la función está representada en la Figura

Exercises.- Graph the functions — **Ejercicios .-** Representar gráficamente las funciones

Page	**Página**
Table	**Tabla**
Figure	**Figura**
Appendix	**Apéndice**

L'Hôpital's rule	**Regla de L'Hôpital**
DH = L'Hôpital's rule.	**DH = Regla de L'Hôpital**
Algebra	**Álgebra**
Limits	**Límites**
Indeterminate forms	**Límites indeterminados**
max = maximum	**max = máximo**
min = minimum	**min = mínimo**

Index	**Índice**
Dictionary	**Diccionario**

English → French

The graph of a function	Le graphique de une fonction
56 functions with solutions + 120 exercises	56 fonctions avec le solutions + 120 exercices
Graphs of polynomial functions	Le graphique d'une fonction polynomiale
Graphs of fractional rational functions	Graphique d'une fonctions rationnelles
Graphs of irrational functions	Graphique d'une fonction irrationnel
Graphs of trigonometric functions	Le graphique d'une fonction trigonométrique
Graphs of exponential and logarithmic functions	Graphique d'une fonction exponentielle et logarithmique
Graphs of absolute value functions	Graphique d'une fonction avec le valeur absolue
Graphs of other functions	Graphique de autres fonction
Graph the function	Représenter graphiquement la fonction
Graph the functions	Représenter graphiquement les fonctions

DOMAIN

DOMAINE DE DÉFINITION

SYMMETRIES AND PERIODICITIES	**symétries et périodicités**
even function	fonction paire
odd function	fonction impaire
periodic function	fonction périodique
period	période
is a even function	est une fonction paire
is a odd function	est une fonction impaire
is a periodic function	est une fonction périodique

SIGN OF FUNCTION

SIGNE DE LA FONCTION

INTERSECTIONS OF THE GRAPH WITH THE COORDINATE AXES

INTERSECTION AVEC LES AXES

LIMITS AND ASYMPTOTES	**LIMITES ET ASYMPTOTES**
vertical asymptote	asymptote verticale
horizontal asymptote	asymptote horizontale
oblique asymptote	asymptote oblique
removable discontinuity	discontinuité réductible
jump discontinuity	discontinuité par saut fini (première espèce)

THE FIRST ORDER DERIVATIVE	**dérivée première**
Derivative	dérivée
Local minimum	minimum local
Local maximum	maximum local
Local minimums and maximums	minimum et maximum locaux
y is increasing	y est croissance
y is decreasing	y est décroissance
THE SECOND ORDER DERIVATIVE	**dérivée seconde**
Inflection point	point d'inflexion
Inflection points	points d'inflexion
y is convex	y est convexe
y is concave	y est concave
The graph of function is represented in Figure	Le graphe de cette fonction est représenté à la Figure
Exercises.- Graph the functions	**Exercices .-** Représenter graphiquement les fonctions
Page	**Page**
Table	**Table**
Figure	**Figure**
Appendix	**Appendice**
L'Hôpital's rule	**Règle de L'Hôpital**
DH = **L'Hôpital's rule.**	*DH* = **Règle de L'Hôpital**
Algebra	**Algèbre**
Limits	**Limites**
Indeterminate forms	**Formes indéterminées**
max = maximum	**max = maximum**
min = minimum	**min = minimum**
Index	**Index**
Dictionary	**Dictionnaire**

English → German

The graph of a function	Der Graph einer funktion
56 functions with solutions + 120 exercises	56 funktionen mit Lösungen + 120 Übungsaufgaben
Graphs of polynomial functions	Graphische Darstellung von Ganzrationale funktionen
Graphs of fractional rational functions	Graphische Darstellung von Gebrochen Rationale funktionen
Graphs of irrational functions	Graphische Darstellung von Irrationale funktionen
Graphs of trigonometric functions	Graphische Darstellung von Trigonometrische funktionen
Graphs of exponential and logarithmic functions	Graphische Darstellung von Exponential- und Logarithmusfunktionen
Graphs of absolute value functions	Graphische Darstellung von Betragsfunktionen
Graphs of other functions	Graphische Darstellung von funktionen
Graph the function	Die funktion ist graphisch darzustellen
Graph the functions	Die funktionen sind graphisch darzustellen

DOMAIN — **DEFINITIONSBEREICH**

SYMMETRIES AND PERIODICITIES — **SYMMETRIE UND PERIODIZITÄT**

even function	gerade funktion
odd function	ungerade funktion
periodic function	periodische funktion
period	periode
is a even function	ist eine gerade funktion
is a odd function	ist eine ungerade funktion
is a periodic function	ist eine periodiche funktion

SIGN OF FUNCTION — **POSITIVITÄT**

INTERSECTIONS OF THE GRAPH WITH THE COORDINATE AXES — **SCHNITTPUNKTE MIT DEN KOORDINATENACHSEN**

LIMITS AND ASYMPTOTES — **GRENZWERTE UND ASYMPTOTEN**

vertical asymptote	vertikale asympote
horizontal asymptote	horizontale asympote

oblique asymptote	schiefe asympote
removable discontinuity	behebbare Unstetigkeit
jump discontinuity	unstetigkeit der ersten Art (Sprung)

THE FIRST ORDER DERIVATIVE	**ERSTE ABLEITUNG**
Derivative	Ableitung
Local minimum	Lokal minimum
Local maximum	Lokal maximum
Local minimums and maximums	Relativen mínimum und maximum
y is increasing	y ist monoton wachsend
y is decreasing	y ist monoton fallend

THE SECOND ORDER DERIVATIVE	**ZWEITE ABLEITUNG**
Inflection point	Wendepunkt
Inflection points	Wendepunkte
y is convex	y ist konvex oder nach oben offen
y is concave	y ist konkaven oder nach unten offen
The graph of function is represented in Figure	Die funktion ist graphisch dargestellt in Bild

Exercises.- Graph the functions	**Aufgaben .-** Representar gráficamente las funciónes

Page	**Pagina**
Table	**Tabelle**
Figure	**Bild**
Appendix	**Anhang**

L'Hôpital's rule	**Regel von L'Hôpital**
DH = **L'Hôpital's rule.**	*DH* = **Regel von L'Hôpital**
Algebra	**Algebra**
Limits	**Grenzwerte**
Indeterminate forms	**Unbestimmte Ausdrücke**
max = maximum	**max = maximum**
min = minimum	**min = minimum**

Index	**Inhalt**
Dictionary	**Wörterbuch**

English → Italian

The graph of a function	Il graphico di una funzione
Graphs of polynomial functions	Graphici di funzioni polinomiali (razionali intere)
Graphs of fractional rational functions	Graphici di funzioni razionali fratte
Graphs of irrational functions	Graphici di funzioni irrazionali
Graphs of trigonometric functions	Graphici di funzioni trigonometriche
Graphs of exponential and logarithmic functions	Graphici di funzioni esponenziali e logaritmiche
Graphs of absolute value functions	Graphici di funzioni in valore assoluto
Graphs of other functions	Graphici di altre funzioni
Graph the function	Rappresentare graphicamente la funzione
Graph the functions	Rappresentare graphicamente le funzioni

DOMAIN — **DOMINIO** (Campo di esistenza)

SYMMETRIES AND PERIODICITIES — **SIMMETRIA E PERIODICITA'**

even function	funzione pari
odd function	funzione dispari
periodic function	funzione periodica
period	periodo
is a even function	è una funzione pari
is a odd function	è una funzioni dispari
is a periodic function	è una funzione periodica

SIGN OF FUNCTION — **SEGNO DELLA FUNZIONE**

INTERSECTIONS OF THE GRAPH WITH THE COORDINATE AXES — **INTERSEZIONI CON GLI ASSI CARTESIANI**

LIMITS AND ASYMPTOTES — **LIMITI E ASINTOTI**

vertical asymptote	asintoto verticale
horizontal asymptote	asintoto orizzontale
oblique asymptote	asintoto obliquo
removable discontinuity	discontinuità eliminabile
jump discontinuity	discontinuità di tipo salto (I specie)

THE FIRST ORDER DERIVATIVE — **DERIVATA PRIMA**

Derivative	Derivata
Local minimum	minimo locale (o relativo)
Local maximum	massimo locale (o relativo)
Local minimums and maximums	Minimo e massimo relativi
y is increasing	y è crescente perché
y is decreasing	y è decrescente

THE SECOND ORDER DERIVATIVE — **DERIVATA SECONDA**

Inflection point	Punto di flesso
Inflection points	Punti di flesso
y is convex	y è convessa o concava verso l'alto
y is concave	y è concava o concava verso il basso
The graph of function is represented in Figure	Il graphico della funzione è rappresentato nella Figura

Exercises.- Graph the functions — **Esercizi .-** Rappresentare graphicamente le funzioni

Page	**Pagina**
Table	**Tabella**
Figure	**Figura**
Appendix	**Appendice**

L'Hôpital's rule	**Regola di De L'Hôpital**
DH = **L'Hôpital's rule.**	*DH* = **Regola di De L'Hôpital.**
Algebra	**Algebra**
Limits	**Limiti**
Indeterminate forms	**Forme indeterminate**
max = maximum	**max = massimo**
min = minimum	**min = minimo**

Index	**Indice**
Dictionary	**Dizionario**

1.- Graph of polynomial functions .

N.1.- Graph the function $y = x^3 - 3x^2 - 3x + 1$.

DOMAIN D

$$D. = R$$

LIMITS AND ASYMPTOTES

$$\lim_{x \to +\infty} x^3 - 3x^2 - 3x + 1 = \lim_{x \to +\infty} x^3 = +\infty$$

$$\lim_{x \to -\infty} x^3 - 3x^2 - 3x + 1 = \lim_{x \to -\infty} x^3 = -\infty$$

SIGN OF FUCTION

$$y \geq 0 \quad \Rightarrow \quad x^3 - 3x^2 - 3x + 1 \quad \Rightarrow \quad (x+1)(x^2 - 4x + 1) \geq 0 \Rightarrow \quad -1 \leq x \leq 2 - \sqrt{3},\ x \geq 2 + \sqrt{3}$$

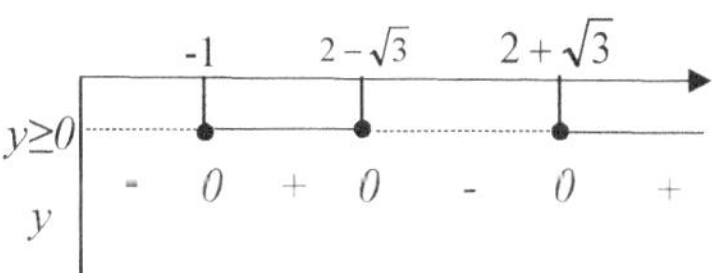

$A\ (-1,0),\ B\left(2 - \sqrt{3}, 0\right),\ C\left(2 + \sqrt{3}, 0\right)$.www.matematicus.com

THE FIRST ORDER DERIVATIVE

$$y' = 3x^2 - 6x - 3 = 3\left(x^2 - 2x - 1\right)$$

$$y' \geq 0 \quad \Rightarrow \quad x^2 - 2x - 1 \geq 0 \Rightarrow x \leq 1 - \sqrt{2},\ x \geq 1 + \sqrt{2} \quad \Rightarrow$$

$$\Rightarrow y \text{ is increasing } \forall x \in \left]-\infty,\ 1 - \sqrt{2}\right] \cup \left[1 + \sqrt{2}, +\infty\right[$$

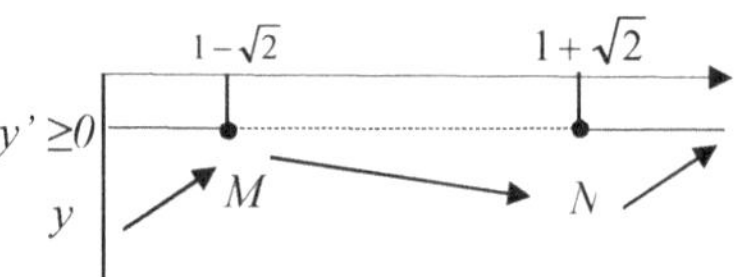

Local minimums and maximums: $M\left(1-\sqrt{2};4\sqrt{2}-4\right)$, $N\left(1+\sqrt{2};-4\sqrt{2}-4\right)$

$$y\left(1-\sqrt{2}\right)=\left(1-\sqrt{2}\right)^3-3\left(1-\sqrt{2}\right)^2-3\left(1-\sqrt{2}\right)+1=$$
$$=1-2\sqrt{2}-3\sqrt{2}+6-3\left(1-2\sqrt{2}+2\right)-3+3\sqrt{2}+1=4\sqrt{2}-4$$

$$y\left(1+\sqrt{2}\right)=-4\sqrt{2}-4$$

THE SECOND ORDER DERIVATIVE

$$y''=6x-6=6(x-1).$$
$$y''\geq 0 \quad\Rightarrow\quad 6(x-1)\geq 0 \Rightarrow x-1\geq 0 \Rightarrow x\geq 1 \quad\Rightarrow$$

$$\Rightarrow y \text{ is convex } \forall x\in\,]1,+\infty[$$

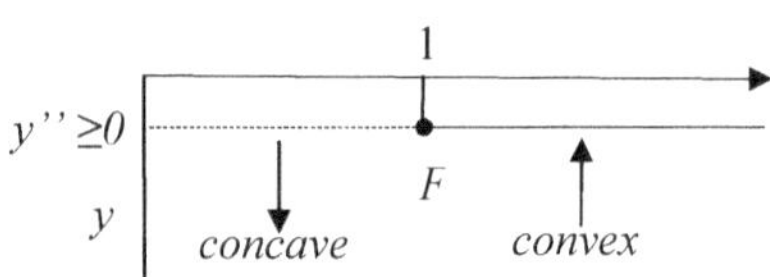

Inflection point : $F(1;-4)$: $\;y(1)=(1)^3-3(1)^2-3(1)+1=-4$.

INTERSECTION OF THE GRAPH WITH THE COORDINATE AXES

$$\begin{cases}x=0\\y=x^3-3x^2-3x+1\end{cases}\Rightarrow\begin{cases}x=0\\y=1\end{cases}\Rightarrow A(0;1)$$

$$\begin{cases}y=0\\y=x^3-3x^2-3x+1\end{cases}\Rightarrow\begin{cases}y=0\\0=(x-1)\left(x^2-4x+1\right)=0\end{cases}\Rightarrow B(-1;0),\,C\left(2-\sqrt{3};0\right),\,D\left(2+\sqrt{3};0\right)$$

The graph of function is represented in Figure 1.1.

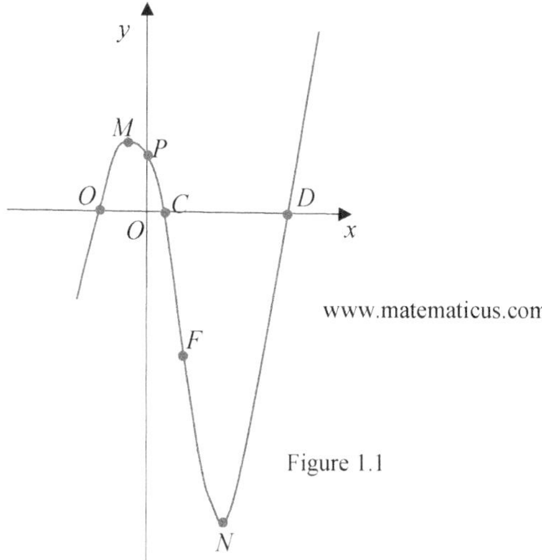

Figure 1.1

N.2.- Graph the function $y = \dfrac{1}{2}x^3 - \dfrac{13}{4}x^2 + 6x - \dfrac{9}{4}$.

DOMAIN D

$$D = R$$

SIGN OF FUNCTION

$$y \geq 0 \quad \Rightarrow \quad \frac{1}{2}x^3 - \frac{13}{4}x + 6x - \frac{9}{4} \geq 0 \Rightarrow \quad 2x^3 - 13x^2 + 24 - 9 \geq 0 \quad \Rightarrow$$

$$\Rightarrow \quad (x-3)^2(2x-1) \geq 0 \quad \Rightarrow \quad x \geq \frac{1}{2}, \; x = 3$$

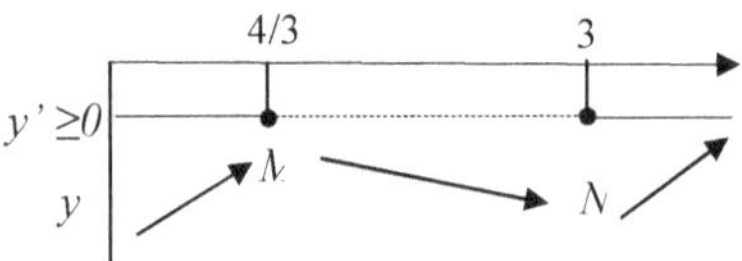

$$B\!\left(\frac{1}{2}, 0\right), O(0,0)$$

THE FIRST ORDER DERIVATIVE

$$y' = \frac{3}{2}x^2 - \frac{13}{2}x + 6.$$

$$y' \geq 0 \quad \Rightarrow \quad \frac{3}{2}x^2 - \frac{13}{2}x + 6 \geq 0 \quad \Rightarrow \quad 3x^2 - 13x + 12 \geq 0 \quad \Rightarrow \quad x \leq \frac{4}{3}, \, x \geq 3$$

Local minimums and maximums: $\quad M\!\left(\dfrac{4}{3}, \dfrac{125}{108}\right), \quad N(3,0)$

$$y\!\left(\frac{4}{3}\right) = \frac{1}{2}\!\left(\frac{4}{3}\right)^3 - \frac{13}{4}\!\left(\frac{4}{3}\right)^2 + 6\!\left(\frac{4}{3}\right) - \frac{9}{4} = \frac{125}{108},$$

$$y(3) = 0 .$$

THE SECOND ORDER DERIVATIVE

$$y' = 3x - \frac{13}{2}.$$

$$y'' \geq 0 \implies 3x - \frac{13}{2} \geq 0 \implies x \geq \frac{13}{6}$$

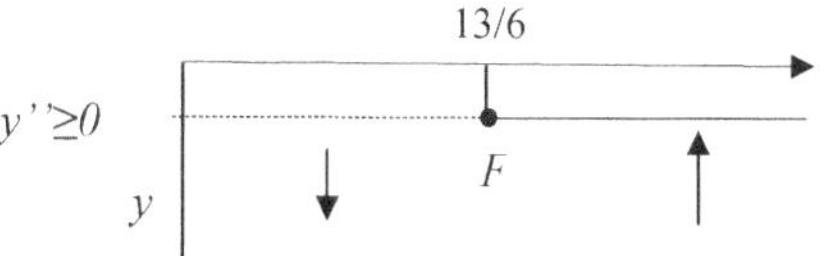

Inflection point $F\left(\frac{13}{6}; \frac{125}{216}\right)$: $y\left(\frac{13}{6}\right) = \frac{1}{2}\left(\frac{13}{6}\right)^3 - \frac{13}{4}\left(\frac{13}{6}\right)^2 + 6\left(\frac{13}{6}\right) - \frac{9}{4} = \frac{125}{216}$.

INTERSECTION OF THE GRAPH WITH THE COORDINATE AXES

$$\begin{cases} y = 0 \\ y = \frac{1}{2}x^3 - \frac{13}{4}x^2 + 6x - \frac{9}{4} \end{cases} \Leftrightarrow \begin{cases} y = 0 \\ (x-3)^2(2x-1) = 0 \end{cases} \implies A(3,0), \ B\left(\frac{1}{2}, 0\right)$$

$$\begin{cases} x = 0 \\ y = \frac{1}{2}x^3 - \frac{13}{4}x^2 + 6x - \frac{9}{4} \end{cases} \implies C\left(0, -\frac{9}{4}\right)$$

The graph of function is represented in Figure 1.2.

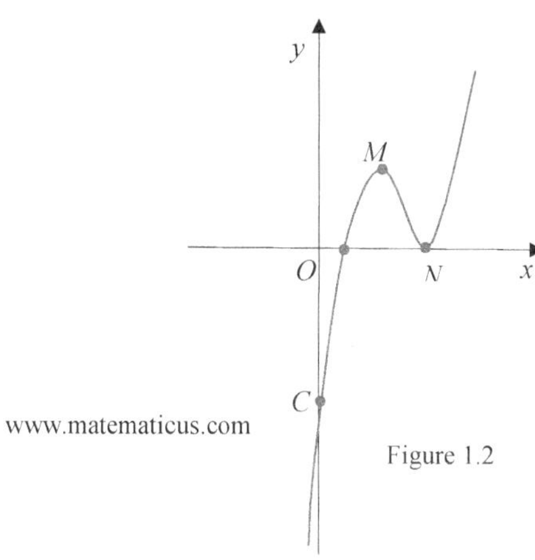

Figure 1.2

N.3.- Graph the function $y = 2x^3 - 3x^2 + 1$.

DOMAIN D

$$D = R$$

SIGN OF FUNCTION

$$y \geq 0 \Rightarrow 2x^3 - 3x^2 + 1 \geq 0 \quad \Rightarrow \quad (x-1)^2(2x+1) \geq 0 \quad \Rightarrow \quad x \geq -\frac{1}{2}, \ x = 1$$

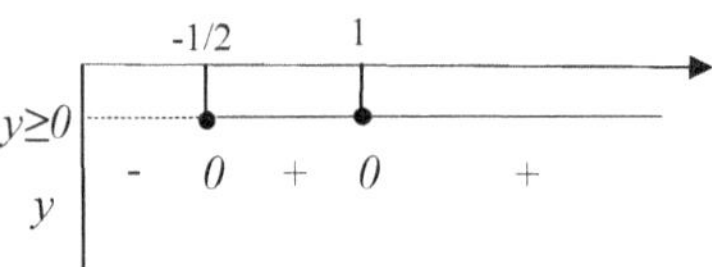

$A(-1/2,0), \ B(1,0)$.

THE FIRST ORDER DERIVATIVE

$y' = 6(x^2 - x)$

$y' \geq 0 \quad \Rightarrow \quad x^2 - x \geq 0 \quad \Rightarrow \quad x \leq 0, \ x \geq 1 \quad \Rightarrow$

$$\Rightarrow \quad y \text{ is increasing } \forall x \in \left]-\infty, \ 0\right] \cup \left[1, +\infty\right[$$

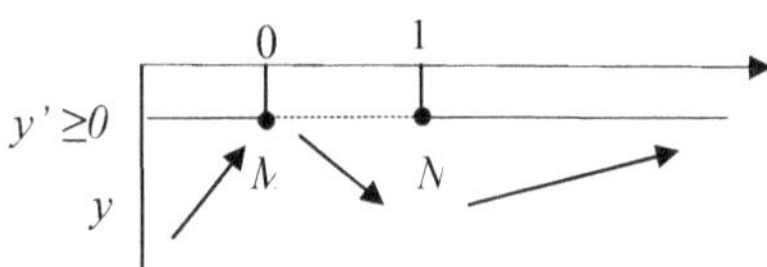

Local minimums and maximums: $M(0;1), N(1;0)$:

$$y(1) = 2(1)^3 - 3(1)^2 + 1 = 2 - 3 + 1 = 0 .$$

THE SECOND ORDER DERIVATIVE

$y'' = 6(2x - 1)$

$$y'' \geq 0 \quad \Rightarrow \quad 6(2x-1) \geq 0 \quad \Rightarrow \quad 2x - 1 \geq 0 \quad \Rightarrow \quad x \geq \frac{1}{2} \quad \Rightarrow$$

$$\Rightarrow y \text{ is convex } \forall x \in \left]\frac{1}{2},+\infty\right[$$

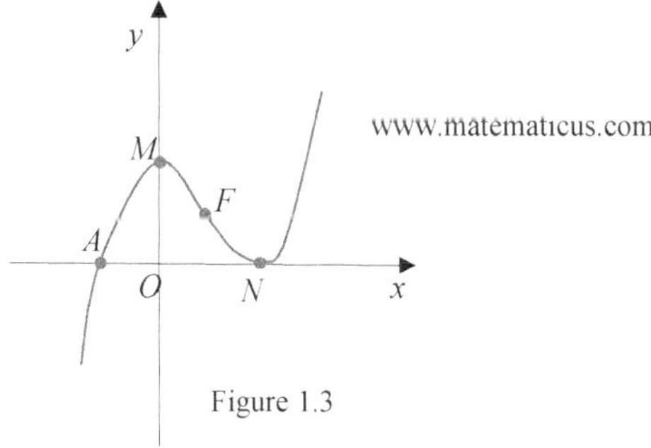

Inflection point $F\left(\frac{1}{2};\frac{1}{2}\right)$:

$$y\left(\frac{1}{2}\right) = 2\left(\frac{1}{2}\right)^3 - 3\left(\frac{1}{2}\right)^2 + 1 = \frac{1}{4} - \frac{3}{4} + 1 = \frac{1}{2}$$

The graph of function is represented in Figure 1.3.

Figure 1.3

N.4.- Graph the function $y = x^4 - x^2$.

DOMAIN D

$$D = R$$

SYMMETRIES AND PERIODICITIES

$$y(-x) = (-x)^4 - (-x)^2 = x^4 - x^2 \implies y \text{ is an even function}.$$

Graph the function $y = x^4 - x^2 \, \forall x \in D_1 = [0, +\infty[$

SIGN OF FUNCTION

$$y \geq 0 \implies x^4 - x^2 \geq 0 \implies x^2(x^2 - 1) \geq 0 \implies x = 0, \quad x \geq 1$$

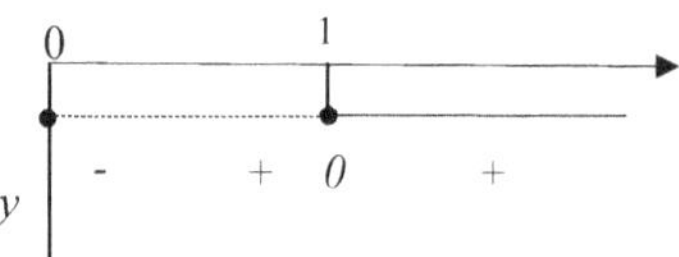

INTERSECTION OF THE GRAPH WITH THE COORDINATE AXES

$$\begin{cases} x = 0 \\ y = x^4 - x^2 \end{cases} \implies O(0,0); \quad \begin{cases} y = 0 \\ y = x^4 - x^2 \end{cases} \implies B(-1;0), \quad C(1,0)$$

THE FIRST ORDER DERIVATIVE

$$y' = 4x^3 - 2x = 2\left(2x^3 - x\right)$$

$$y' \geq 0 \implies 2\left(2x^3 - x\right) \geq 0 \implies 2[x(2x^2 - 1)] \geq 0 \implies x = 0, \, x \geq \sqrt{\frac{1}{2}}$$

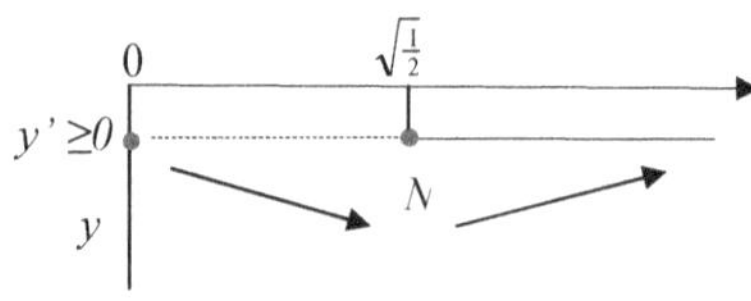

Local minimums and maximums: $M(0;0)$, $N\left(\sqrt{\dfrac{1}{2}};-\dfrac{1}{4}\right)$.

$$y(0)=0, \qquad y\left(\sqrt{\dfrac{1}{2}}\right)=\left(\sqrt{\dfrac{1}{2}}\right)^{4}-\left(\sqrt{\dfrac{1}{2}}\right)^{2}=\dfrac{1}{4}-\dfrac{1}{2}=-\dfrac{1}{4}$$

THE SECOND ORDER DERIVATIVE

$$y''=12x^{2}-2=2\left(6x^{2}-1\right)$$

$$y''\geq 0 \quad \Rightarrow \quad 6x^{2}-1\geq 0 \quad \Rightarrow \quad x\geq\sqrt{\dfrac{1}{6}} \qquad \Rightarrow y \text{ is convex } \forall x\in\left]\sqrt{\dfrac{1}{6}},+\infty\right[$$

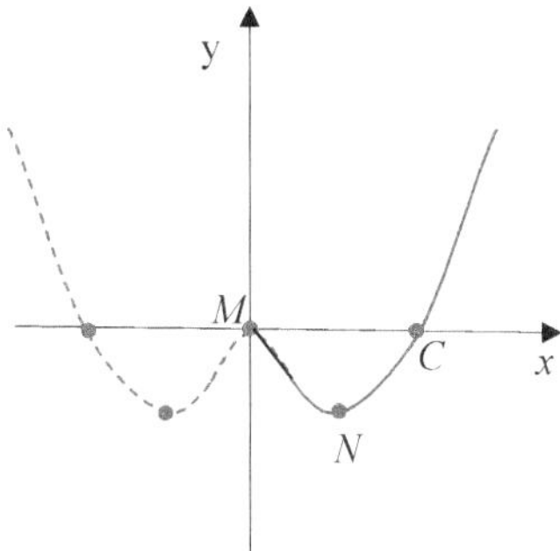

Inflection point $F\left(\sqrt{\dfrac{1}{6}};-\dfrac{5}{36}\right)$: $y\left(\sqrt{\dfrac{1}{6}}\right)=\left(\sqrt{\dfrac{1}{6}}\right)^{4}-\left(\sqrt{\dfrac{1}{6}}\right)^{2}=\dfrac{1}{36}-\dfrac{1}{6}=-\dfrac{5}{36}$.

The graph of function is represented in Figure 1.4,

Figure 1.4

N.5.- Graph the function $y = \begin{cases} -x^3 - 2x^2 & \forall\ x \le 1 \\ -4x^2 + 16x - 15 & \forall\ x > 1 \end{cases}$.

DOMAIN D

$$D = R$$

$$y = y_1 \cup y_2$$

$$y_1 = -x^3 - 2x^2 \quad \forall x \in \left]-\infty, 1\right] \quad \cup \quad y_2 = -4x^2 + 16x - 15 \quad \forall x \in \left]1, +\infty\right[$$

Graph the function $y_1 = -x^3 - 2x^2 \quad \forall x \in \left]-\infty, 1\right]$

SIGN OF FUNCTION

$$y_1 \ge 0 \quad \Rightarrow \quad -x^3 - 2x^2 \ge 0 \quad \Rightarrow \quad x^2(-x - 2) \ge 0 \quad \Rightarrow \quad x = 0, \quad x \le -2$$

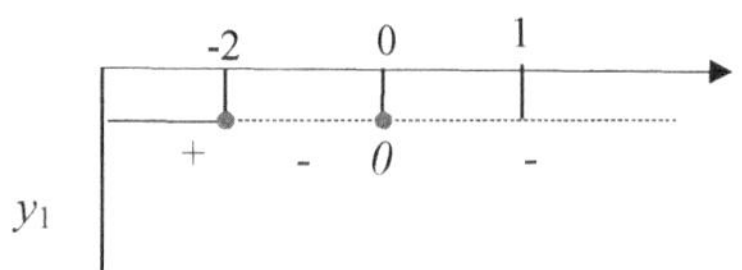

INTERSECTION OF THE GRAPH WITH THE COORDINATE AXES

$O(0;0)$ e $R(-2;0)$.

THE FIRST ORDER DERIVATIVE

$y_1' = -3x^2 - 4x.$

$$y_1' \ge 0 \quad \Rightarrow \quad -3x^2 - 4x \ge 0 \quad \Rightarrow \quad -\frac{4}{3} \le x \le 0$$

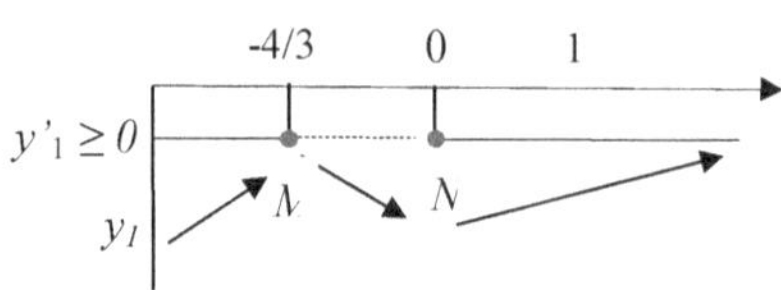

Local minimums and maximums: $M(0;0)$, $N\left(-\dfrac{4}{3};-\dfrac{32}{27}\right)$

$$y_1\left(-\frac{4}{3}\right)=-\left(-\frac{4}{3}\right)^3-2\left(-\frac{4}{3}\right)^2=-\left(-\frac{64}{27}\right)-2\left(\frac{16}{9}\right)=-\frac{32}{27}.$$

THE SECOND ORDER DERIVATIVE

$y_1''=-6x-4.$

$y_1''\geq 0 \quad\Rightarrow\quad -6x-4\geq 0 \quad\Rightarrow\quad x\leq -\dfrac{2}{3}$

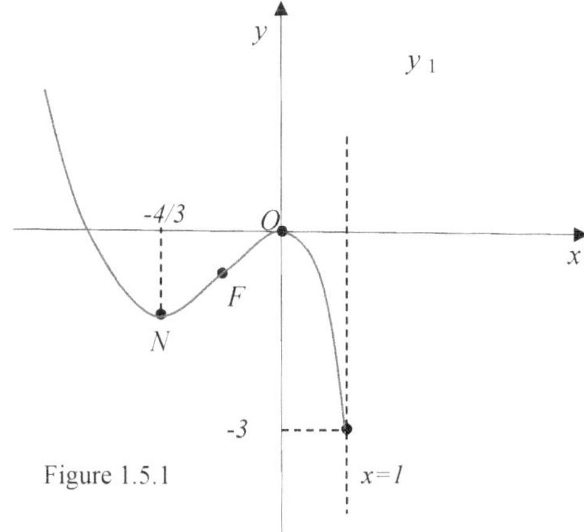

Inflection point: $F\left(-\dfrac{2}{3};-\dfrac{16}{27}\right)$: $y_1\left(-\dfrac{2}{3}\right)=-\left(-\dfrac{2}{3}\right)^3-2\left(-\dfrac{2}{3}\right)^2=\dfrac{8}{27}-\dfrac{8}{9}=-\dfrac{16}{27}.$

The graph of function y_1 is represented in Figure 1.5.1

Figure 1.5.1

Graph the function $y_2 = -4x^2 + 16x - 15 \quad \forall x \in\ ]1,+\infty[$

The graph of function y_2 is represented in Figure 1.5.2

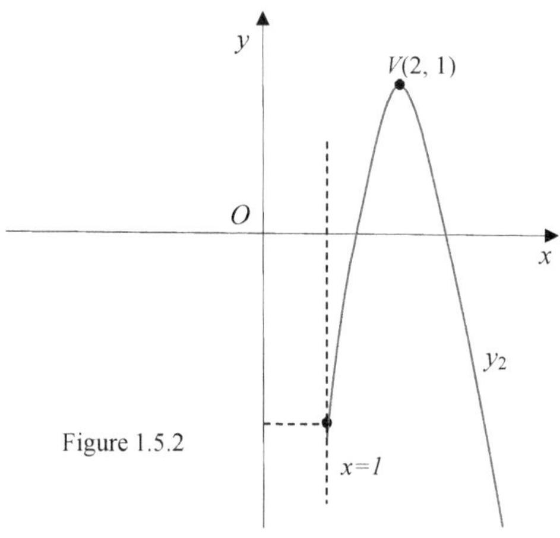

Figure 1.5.2

The graph of function $y = y_1 \cup y_2$ is represented in Figure 1.5.

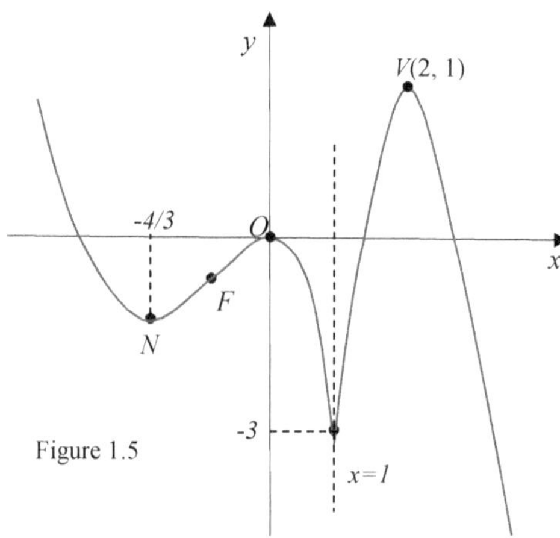

Figure 1.5

N.6.- Graph the function $y = \begin{cases} -x & x \le 1 \\ 4 & 1 < x < 3 \\ x-3 & x \ge 3 \end{cases}$.

DOMAIN D

$$D = R$$

$$y = y_1 \cup y_2 \cup y_3$$

$y_1 = -x \qquad \forall x : x \le 1$

$y_2 = 4 \qquad \forall x : \ 1 < x < 3$

$y_3 = x - 3 \quad \forall x : \ x \ge 3$

The graph of function is represented in Figure1.6.

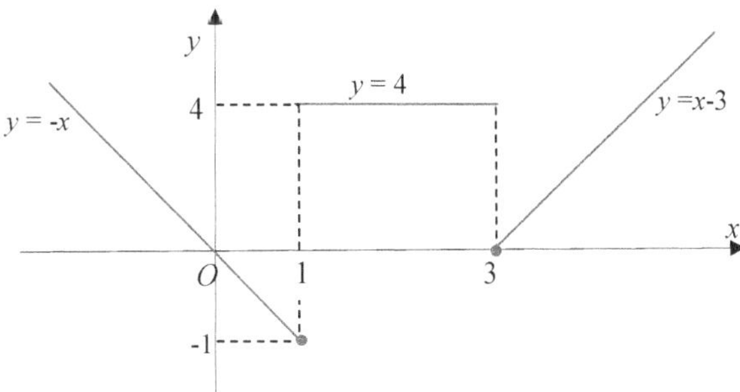

Figure 1.6

Exercises.- Graph the functions.

a) $y = x^3 (x^2 - 1)$;   b) $y = \begin{cases} x^2 - 1 & \forall x \ x > 0 \\ -x^2 - 4x + 1 & \forall x \ x \leq 0 \end{cases}$;

c) $y = \begin{cases} 2 & \forall x \ 0 \leq x \leq 2 \\ 4x - 2 & \forall x \ x > 2 \\ -x^2 - 2x + 2 & \forall x \ x < 0 \end{cases}$;

d) $y = x^4 - 6x^2 + 5$;   e) $y = -x^4 - 13x^2 + 36$;   f) $y = (x^2 + 1)(x^2 - 3x + 4)$;

g) $y = (x-1)(x^2 - 4)$;   h) $y = x^3 (x^2 - 1)$;   i) $y = \dfrac{1}{6}x^3 - \dfrac{1}{4}x^2 - x + \dfrac{1}{12}$;

l) $y = x^5 + \dfrac{335}{88}x^4 - \dfrac{175}{33}x^3 - \dfrac{115}{4}x^2 - \dfrac{255}{11}x + \dfrac{13841}{264}$.

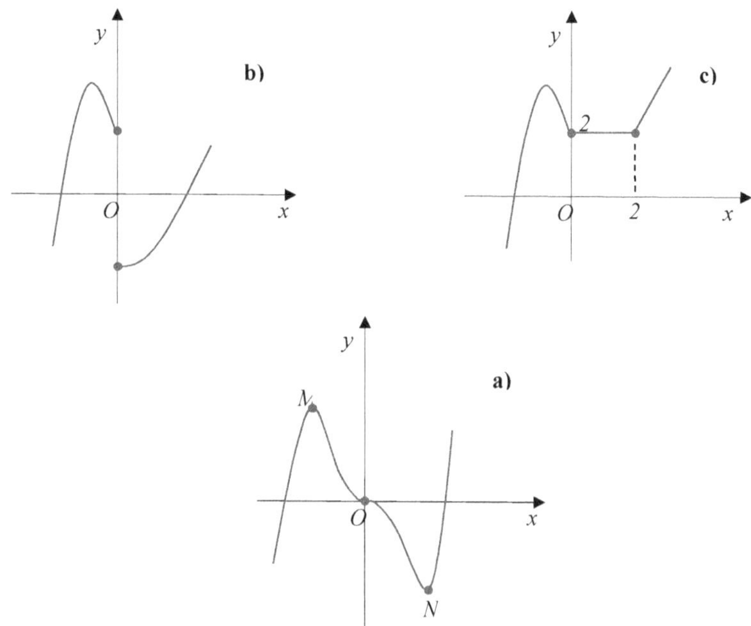

$M\left(-\sqrt{3/5};0.2\right), N\left(\sqrt{3/5};-0,2\right)$ www.matematicus.com

2.- Graph of fractional rational functions.

N.1.- Graph the function $y = \dfrac{x^2 + 1}{x^2 - 1}$.

DOMAIN D

$$x^2 - 1 \neq 0 \quad \Rightarrow \quad x \neq \pm 1 \quad \Rightarrow \quad D = R - \{\pm 1\}$$

SYMMETRIES AND PERIODICITIES

$$y(-x) = \frac{(-x)^2 + 1}{(-x)^2 - 1} = \frac{x^2 + 1}{x^2 - 1} = y(x) \quad \Rightarrow \quad y \text{ is a even function}$$

Graph the function $y = \dfrac{x^2 + 1}{x^2 - 1} \quad \forall x \in D_1 = [0, +\infty[- \{1\}$

SIGN OF FUNCTION

$$y \geq 0 \quad \Rightarrow \quad \frac{x^2 + 1}{x^2 - 1} \geq 0 \quad \Rightarrow \quad x^2 - 1 > 0 \quad \Rightarrow x > 1$$

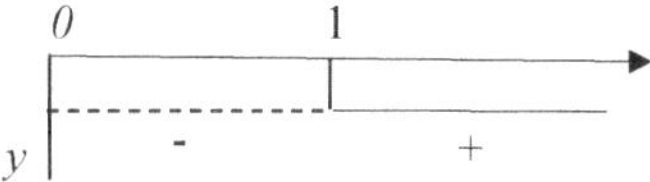

LIMITS AND ASYMPTOTES [1]

$$\lim_{x \to 1^+} \frac{x^2 + 1}{x^2 - 1} = \frac{2}{0^+} = +\infty$$

$$\Rightarrow \quad x = 1 \text{ vertical asymptote}$$

$$\lim_{x \to 1^-} \frac{x^2 + 1}{x^2 - 1} = \frac{2}{0^-} = -\infty$$

$$\lim_{x \to \pm\infty} \frac{x^2 + 1}{x^2 - 1} = 1 \quad \Rightarrow \quad y = 1 \text{ horizontal asymptote}$$

---

[1] Appendix.

INTERSECTION OF THE GRAPH WITH THE COORDINATE AXES

$$\begin{cases} x = 0 \\ y = \dfrac{x^2+1}{x^2-1} \end{cases} \quad \Rightarrow \quad A(0;-1)$$

THE FIRST ORDER DERIVATIVE

$$y' = \frac{2x(x^2-1)-(x^2+1)2x}{(x^2-1)^2} = \frac{-4x}{(x^2-1)^2}$$

$\Rightarrow \ y$ is decreasing $\forall x \in [0,+\infty[$

$$y' \geq 0 \quad \Rightarrow \quad \frac{-4x}{(x^2-1)^2} \geq 0 \quad \Rightarrow \quad x = 0$$

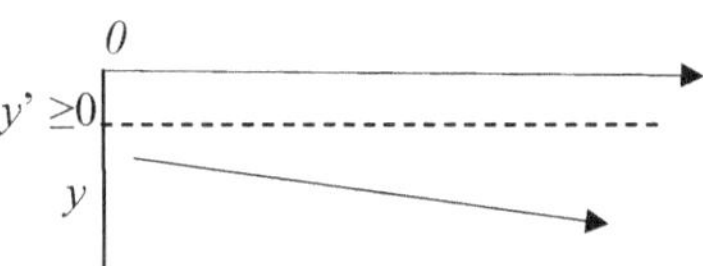

Local maximum: $M(0,-1)$.

THE SECOND ORDER DERIVATIVE

$$y'' = \frac{-4(x^2-1)^2 - \left[-4x - 2(x^2-1)2x\right]}{(x^2-1)^4} = \frac{-4(x^2-1)^2 + 16x^2(x^2-1)}{(x-1)^4} = \frac{4(3x^2+1)}{(x^2-1)^3}$$

$$y'' \geq 0 \Rightarrow \quad \frac{4(3x^2+1)}{(x^2-1)^3} \geq 0 \quad \Rightarrow \quad x^2-1 > 0 \quad \Rightarrow \quad x > 1$$

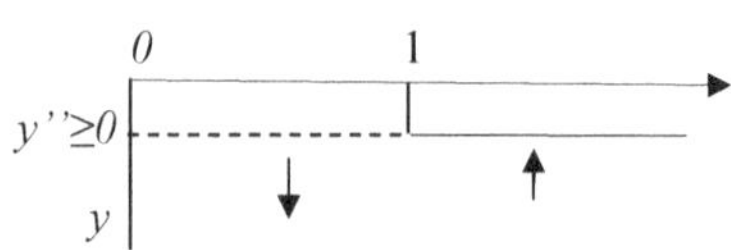

The graph of function is represented in Figure2.1.

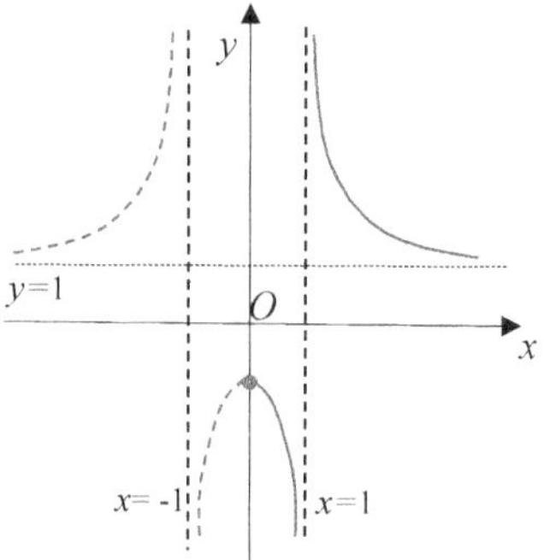

Figure 2.1

N.2.- Graph the function $\quad y = \dfrac{1+x^3}{x^2}$

DOMAIN *D*

$$x^2 \neq 0 \quad \Rightarrow \quad x \neq 0 \quad \Rightarrow \quad D = R - \{0\}$$

SIGN OF FUNCTION

$$y \geq 0 \quad \Rightarrow \quad \frac{1+x^3}{x^2} \geq 0 \quad \Rightarrow \quad 1+x^3 \geq 0 \quad \Rightarrow \quad (1+x)(1-x+x^2) \geq 0 \quad \Rightarrow \quad 1+x \geq 0 \quad \Rightarrow$$

$$\Rightarrow \quad x \geq -1$$

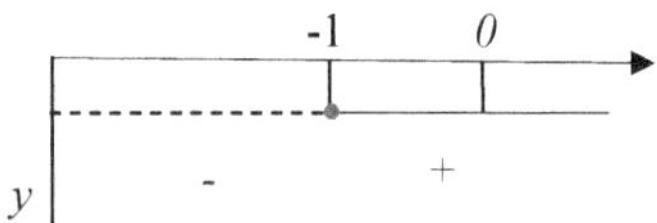

LIMITS AND ASYMPTOTES

$$\lim_{x \to 0^+} \frac{1+x^3}{x^2} = \frac{1}{0^+} = +\infty$$

$$\lim_{x \to 0^-} \frac{1+x^3}{x^2} = \frac{1}{(0^-)^2} = \frac{1}{0^+} = +\infty$$

$\Rightarrow x = 0$ vertical asymptote

$$\lim_{x \to \pm\infty} \frac{1+x^3}{x^2} = \pm\infty \quad \Rightarrow \quad y = mx + n. \text{ oblique asymptote:}$$

$$m = \lim_{x \to \pm\infty}\left(\frac{1+x^3}{x^2} \times \frac{1}{x}\right) = 1, \quad n = \lim_{x \to \pm\infty}\left(\frac{1+x^3}{x^2} - x\right) = \lim_{x \to \pm\infty}\left(\frac{1}{x^2}\right) = 0 \quad \Rightarrow$$

$$\Rightarrow \quad y = x \quad \text{oblique asymptote}$$

INTERSECTION OF THE GRAPH WITH THE COORDINATE AXES

$$\begin{cases} y = 0 \\ y = \dfrac{1+x^3}{x^2} \end{cases} \Rightarrow \begin{cases} y = 0 \\ 0 = \dfrac{1+x^3}{x^2} \end{cases} \Rightarrow \begin{cases} y = 0 \\ 0 = 1+x^3 \end{cases} \Rightarrow \begin{cases} y = 0 \\ -1 = x^3 \end{cases} \Rightarrow \begin{cases} y = 0 \\ -1 = x \end{cases} \Rightarrow A(-1;0)$$

THE FIRST ORDER DERIVATIVE

$$y' = \frac{3x^2 \, x^2 - \left(1 + x^3\right)2x}{x^4} = \frac{x^3 - 2}{x^3}$$

$$y' \geq 0 \quad \Rightarrow \quad \frac{x^3 - 2}{x^3} \geq 0 \quad \Rightarrow \quad x < 0 \quad \vee \quad x \geq \sqrt[3]{2}$$

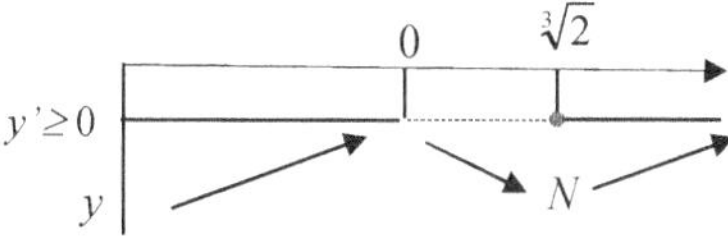

Local minimum $N\left(\sqrt[3]{2}\,; \frac{3}{2}\sqrt[3]{2}\right)$: $\quad y\left(\sqrt[3]{2}\right) = \frac{1 + \left(\sqrt[3]{2}\right)^3}{\left(\sqrt[3]{2}\right)^2} = \frac{1 + 2}{\sqrt[3]{4}} = \frac{3}{\sqrt[3]{4}}\frac{\sqrt[3]{2}}{\sqrt[3]{2}} = \frac{3}{2}\sqrt[3]{2}.$

THE SECOND ORDER DERIVATIVE

$$y'' = \frac{6}{x^4} \quad \Rightarrow \quad y'' > 0 \, \forall x \in D \quad \Rightarrow \quad y \text{ is convex } \forall x \in D$$

The graph of function is represented in Figure 2.2

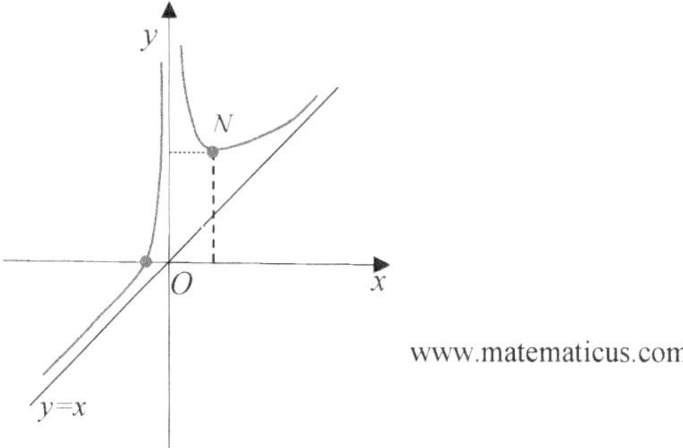

www.matematicus.com

Figure 2.2

N.3.- Graph the function $y = \dfrac{x^3}{2x^2 - 1}$.

DOMAIN D

$$2x^2 - 1 \neq 0 \Rightarrow x \neq \pm\sqrt{\frac{1}{2}} = \pm\frac{\sqrt{2}}{2} \quad \Rightarrow \quad D = R - \left\{\pm\frac{\sqrt{2}}{2}\right\}$$

SYMMETRIES AND PERIODICITIES

$$y = \frac{(-x)^3}{2(-x)^2 - 1} = -\frac{x^3}{2x^2 - 1} = -y \quad \Rightarrow \quad y \text{ is a odd function}$$

Graph the function $\quad y = \dfrac{x^3}{2x^2 - 1} \quad \forall x \in D_1 = \left[0, +\infty\right[- \left\{\frac{\sqrt{2}}{2}\right\}$

SIGN OF FUNCTION

$$y \geq 0 \quad \Rightarrow \quad \frac{x^3}{2x^2 - 1} \geq 0 \quad \Rightarrow \quad x = 0 \quad \vee \quad x > \frac{\sqrt{2}}{2}$$

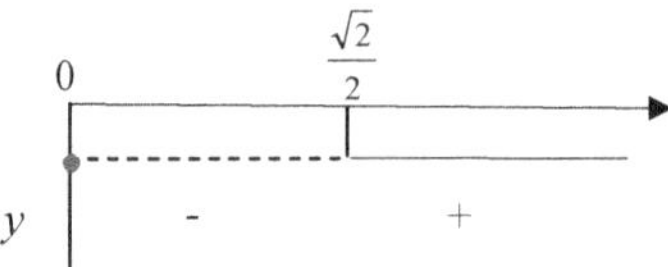

LIMITS AND ASYMPTOTES

$$\lim_{x \to \frac{\sqrt{2}}{2}^+} \frac{x^3}{2x^2 - 1} = \frac{\sqrt[3]{8}/8}{0^+} = +\infty$$

$$\lim_{x \to \frac{\sqrt{2}}{2}^-} \frac{x^3}{2x^2 - 1} = \frac{\sqrt[3]{8}}{0^-} = -\infty$$

$$\Rightarrow \quad x = \frac{\sqrt{2}}{2} \quad \text{vertical asymptote.}$$

$$\lim_{x \to \pm\infty} \frac{x^3}{2x^2 - 1} = \pm\infty ,$$

$$m = \lim_{x \to \pm\infty}\left(\frac{x^3}{2x^2 - 1} \times \frac{1}{x}\right) = \frac{1}{2}, \quad n = \lim_{x \to \pm\infty}\frac{x^3}{2x^2 - 1} - \frac{1}{2}x = 0 \quad \Rightarrow$$

$$\Rightarrow \quad y = x/2 \quad \text{oblique asymptote}$$

THE FIRST ORDER DERIVATIVE

$$y' = \frac{3x^2\left(2x^2 - 1\right) - x^3\, 4x}{\left(2x - 1\right)^2} = \frac{x^2\left(2x^2 - 3\right)}{\left(2x^2 - 1\right)^2}.$$

$$y' \geq 0 \Rightarrow \frac{x^2\left(2x^2 - 3\right)}{\left(2x^2 - 1\right)^2} \geq 0 \Rightarrow x^2\left(2x^2 - 3\right) \geq 0 \quad \Rightarrow \quad x = 0 \quad \vee \quad x \geq \frac{\sqrt{6}}{2}$$

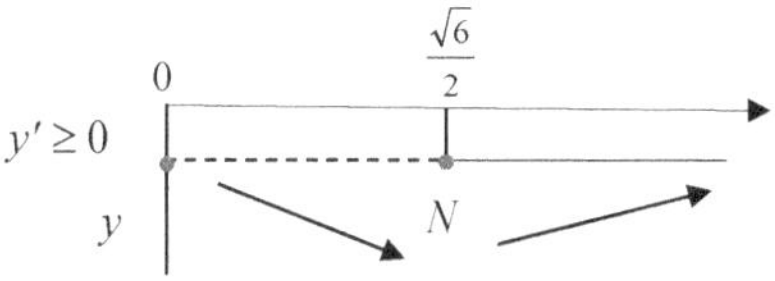

Local minimum: $N\left(\dfrac{\sqrt{6}}{2}; \dfrac{3}{8}\sqrt{6}\right),$ $\quad y\left(\dfrac{\sqrt{6}}{2}\right) = \dfrac{\left(\dfrac{\sqrt{6}}{2}\right)^3}{2\left(\dfrac{\sqrt{6}}{2}\right)^2 - 1} = \dfrac{\dfrac{\left(\sqrt{6}\right)^3}{8}}{3 - 1} = \dfrac{1}{8}6\sqrt{6}\,\dfrac{1}{2} = \dfrac{3}{8}\sqrt{6}$

Inflection point $x = 0 \Rightarrow F(0,0)$.

THE SECOND ORDER DERIVATIVE

$$y'' = \frac{2x\left(2x^2 - 3\right) + x^2\, 4x - x^2\left(2x^2 - 3\right)\left[2\left(2x^2 - 1\right)4x\right]}{\left(2x^2 - 1\right)^4} = \frac{2x\left(2x^2 + 3\right)}{\left(2x^2 - 1\right)^3}.$$

$$y'' \geq 0 \quad \Rightarrow \quad \frac{2x\left(2x^2 + 3\right)}{\left(2x^2 - 1\right)^3} \geq 0 \quad \Rightarrow \quad x = 0 \quad \vee \quad x > \frac{\sqrt{2}}{2}$$

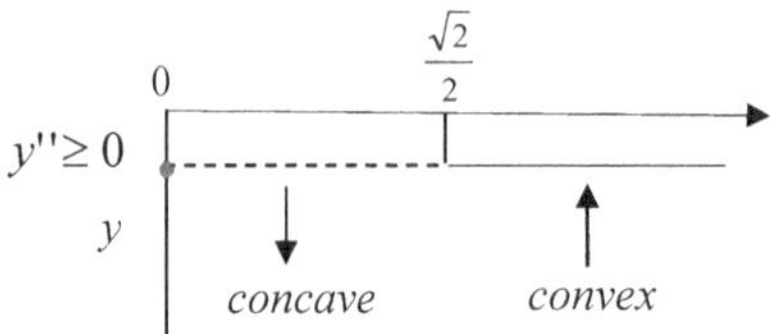

Inflection point $F(0,0)$.

The graph of function is represented in Figure 2.3.

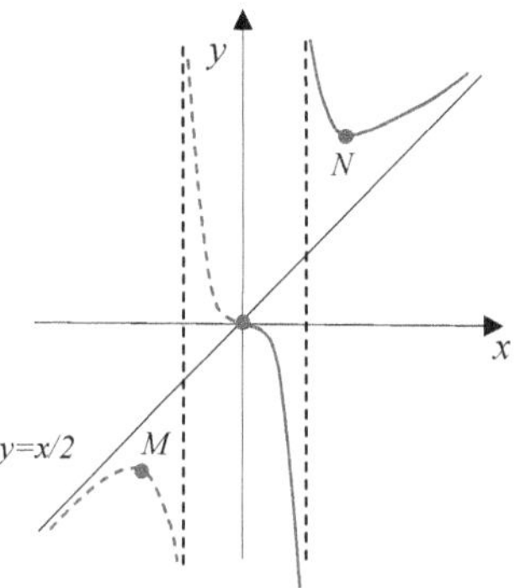

Figure 2.3

$$M\left(-\frac{\sqrt{6}}{2};-\frac{3}{8}\sqrt{6}\right)$$

www.matematicus.com

N.4.- Graph the function $y = \dfrac{2x-1}{2x^3}$.

DOMAIN D

$$2x^3 \neq 0 \implies x^3 \neq 0 \implies x \neq 0 \implies D = R - \{0\}.$$

SIGN OF FUNCTION

$$y \geq 0 \implies \frac{2x-1}{2x^3} \geq 0 \implies x < 0 \quad \vee \quad x \geq \frac{1}{2}$$

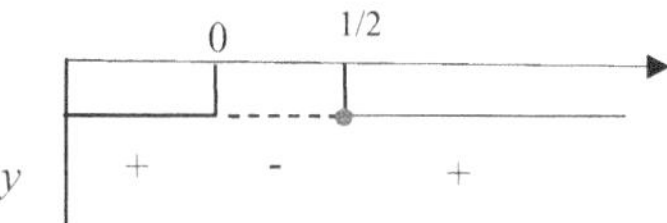

$A\,(1/2, 0)$

LIMITS AND ASYMPTOTES

$$\lim_{x \to 0^-} \frac{2x-1}{2x^3} = \frac{-1}{0+} = -\infty$$

$$\lim_{x \to 0^+} \frac{2x-1}{2x^3} = \frac{-1}{0^-} = +\infty$$

$\implies x = 0$ vertical asymptote.

$$\lim_{x \to \pm\infty} \frac{2x-1}{2x^3} = 0 \implies y = 0 \text{ horizontal asymptote}$$

INTERSECTION OF THE GRAPH WITH THE COORDINATE AXES

$$\begin{cases} y = 0 \\ y = \dfrac{2x-1}{2x^3} \end{cases} \implies \begin{cases} y = 0 \\ 0 = \dfrac{2x-1}{2x^3} \end{cases} \implies \begin{cases} y = 0 \\ 0 = 2x-1 \end{cases} \implies \begin{cases} y = 0 \\ x = \dfrac{1}{2} \end{cases} \implies A(1/2;0)$$

THE FIRST ORDER DERIVATIVE

$$y' = \frac{3 - 4x}{2x^4}$$

$$y' \geq 0 \implies \frac{3-4x}{2x^4} \geq 0 \implies 3 - 4x \geq 0 \implies x \leq 3/4$$

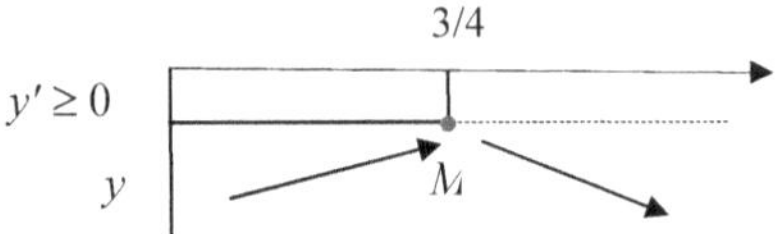

Local maximum: $M(3/4;\ 16/27)$

THE SECOND ORDER DERIVATIVE

$$y'' = \frac{6(x-1)}{x^5}$$

$$y'' \geq 0 \quad \Rightarrow \quad \frac{6(x-1)}{x^5} \geq 0 \quad \Rightarrow \quad x < 0 \quad \vee \quad x \geq 1$$

Inflection point: $F(1;1/2)$

$$y(1) = \frac{2(1)-1}{2(1)^3} = \frac{2-1}{2} = \frac{1}{2}$$

The graph of function is represented in Figure 2.4.

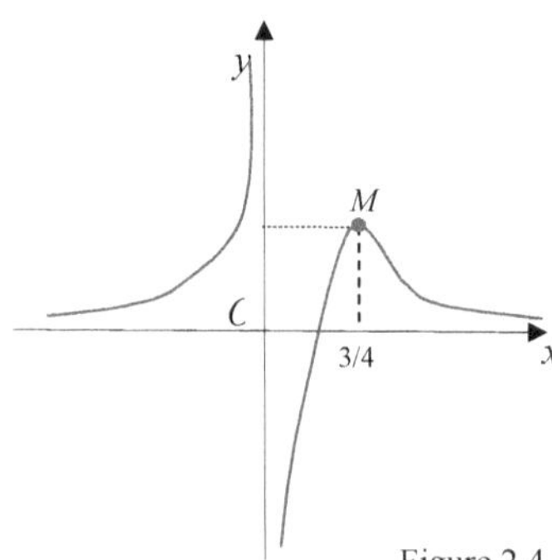

Figure 2.4

N. 5.- Graph the function $y = \dfrac{6x^2 + 2x + 3}{2(2x^2 + 1)}$.

DOMAIN D

$$2(2x^2 + 1) \neq 0 \quad \forall x \in R \quad \Rightarrow \quad D = R$$

SIGN OF FUNCTION

$$y > 0 \quad \forall x \in R$$

LIMITS AND ASYMPTOTES

$$\lim_{x \to \pm\infty} \frac{6x^2 + 2x + 3}{2\left(2x^2 + 1\right)} = \frac{3}{2} \quad \Rightarrow \quad y = \frac{3}{2} \ \text{ horizontal asymptote}$$

INTERSECTION OF THE GRAPH WITH THE COORDINATE AXES

$$\begin{cases} x = 0 \\ y = \dfrac{6x^2 + 2x + 3}{2(2x^2 + 1)} \end{cases} \Rightarrow \quad A\left(0; \frac{3}{2}\right)$$

THE FIRST ORDER DERIVATIVE

$$y' = \frac{(12x + 2)\left[2(2x^2 + 1)\right] - 8x(6x^2 + 2x + 3)}{\left[2(2x^2 + 1)\right]^2} = \frac{1 - 2x^2}{\left(2x^2 + 1\right)^2}.$$

$$y' \geq 0 \Rightarrow \quad \frac{1 - 2x^2}{\left(2x^2 + 1\right)^2} \geq 0 \quad \Rightarrow \quad 1 - 2x^2 \geq 0 \quad \Rightarrow \quad -\frac{\sqrt{2}}{2} \leq x \leq \frac{\sqrt{2}}{2}$$

Local minimums and maximums: $\ M\left(\dfrac{\sqrt{2}}{2}; \dfrac{6 + \sqrt{2}}{4}\right), N\left(-\dfrac{\sqrt{2}}{2}; \dfrac{6 - \sqrt{2}}{4}\right).$

$$y\left(\frac{\sqrt{2}}{2}\right) = \frac{6\left(\frac{\sqrt{2}}{2}\right)^2 + 2\left(\frac{\sqrt{2}}{2}\right) + 3}{2\left(2\left(\frac{\sqrt{2}}{2}\right)^2 + 1\right)} = \frac{6\frac{1}{2} + \sqrt{2} + 3}{2\left(2\frac{1}{2} + 1\right)} = \frac{6 + \sqrt{2}}{4}$$

THE SECOND ORDER DERIVATIVE

$$y'' = \frac{4x(2x^2 - 3)}{(2x^2 + 1)^3}.$$

$$y'' \geq 0 \quad \Rightarrow \quad \frac{4x(2x^2 - 3)}{(2x^2 + 1)^3} \geq 0 \quad \Rightarrow \quad x(2x^2 - 3) \geq 0 \quad \Rightarrow \quad -\sqrt{\frac{3}{2}} \leq x \leq 0 \quad \vee \quad x \geq \sqrt{\frac{3}{2}}$$

Inflection points: $F_1\left(-\sqrt{\frac{3}{2}}; \frac{12 - \sqrt{6}}{8}\right)$, $F_2\left(0; -\frac{3}{2}\right)$, $F_3\left(\sqrt{\frac{3}{2}}; \frac{12 + \sqrt{6}}{8}\right)$.

The graph of function is represented in fig 2.5.

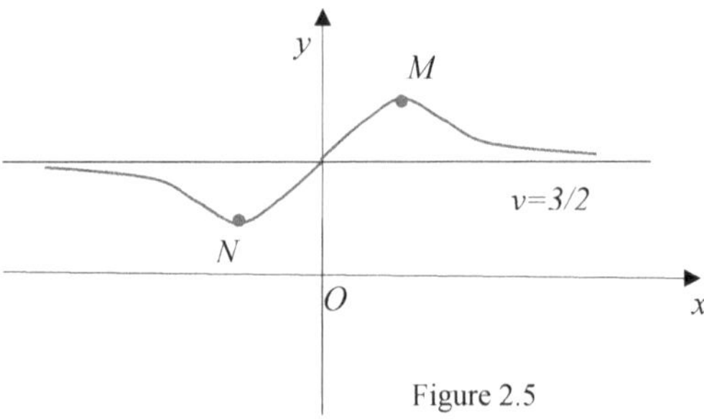

Figure 2.5

N.6.- Graph the function $y = \begin{cases} \dfrac{1}{2x+1} & \forall x : x \le 0 \\[3mm] 1 & \forall x : 0 < x \le \dfrac{1}{2} \\[3mm] \dfrac{x(1-2x)}{2x+1} & \forall x : x > \dfrac{1}{2} \end{cases}$

DOMAIN D

$$2x + 1 \ne 0 \quad \Rightarrow \quad x \ne -1/2 \quad \Rightarrow \quad D = R - \left\{\dfrac{1}{2}\right\}$$

$$y = y_1 \cup y_2 \cup y_3$$

$$y_1 = \dfrac{1}{2x+1} \qquad \forall x: x \le 0,$$

$$y_2 = 1, \qquad \forall x \in \,]0,1/2],$$

$$y_3 = \dfrac{x(1-2x)}{2x+1} \qquad \forall x: x > 1/2.$$

Graph the function $y_1 = \dfrac{1}{2x+1} \quad \forall x : x \le 0$

DOMAIN

$$2x + 1 \ne 0 \iff x \ne -1/2 \quad \Rightarrow \quad D_1 = \,]-\infty,0] - \{-1/2\}$$

SIGN OF FUNCTION

$$y_1 \ge 0 \quad \Rightarrow \quad \begin{cases} \dfrac{1}{2x+1} \ge 0 \\[3mm] x \le 0 \end{cases} \Rightarrow \begin{cases} 2x + 1 > 0 \\[2mm] x \le 0 \end{cases} \Rightarrow \quad -\dfrac{1}{2} < x \le 0$$

LIMITS AND ASYMPTOTES

$$\begin{cases} \lim\limits_{x\to-\frac{1}{2}^+} y = \lim\limits_{x\to-\frac{1}{2}^+} \dfrac{1}{2x+1} = +\infty \\[2em] \lim\limits_{x\to-\frac{1}{2}^-} y = \lim\limits_{x\to-\frac{1}{2}^-} \dfrac{1}{2x+1} = -\infty \end{cases} \Rightarrow x = -\frac{1}{2} \quad \text{vertical asymptote}$$

$$\lim\limits_{x\to-\infty} y = \lim\limits_{x\to-\infty} \dfrac{1}{2x+1} = 0 \Rightarrow y = 0 \quad \text{horizontal asymptote } (x\to-\infty)$$

THE FIRST ORDER DERIVATIVE

$$y' = \frac{-2}{\left(2x+1\right)^2}$$

$$y' \geq 0 \quad \Rightarrow \quad = \frac{-2}{\left(2x+1\right)^2} \geq 0 \Rightarrow \forall x \in \varnothing \cap D_1 \quad \Rightarrow y \text{ is decreasing } \forall x \in D_1$$

The graph of function y_1 is represented in Figure 2.6.1.

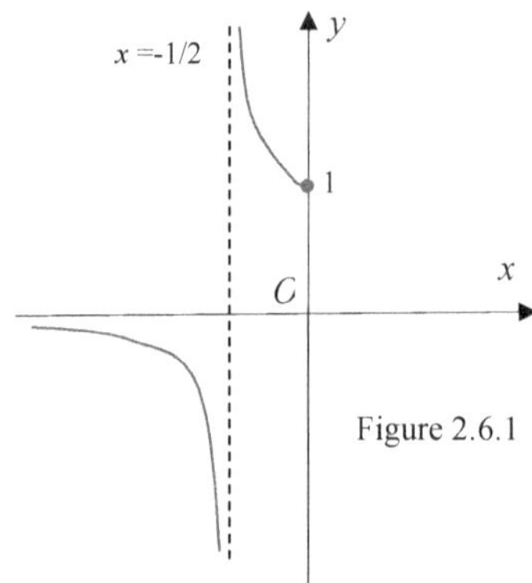

Figure 2.6.1

www.matematicus.com

Graph the function $y_2 = 1 \quad \forall x \in \,]0,1/2]$

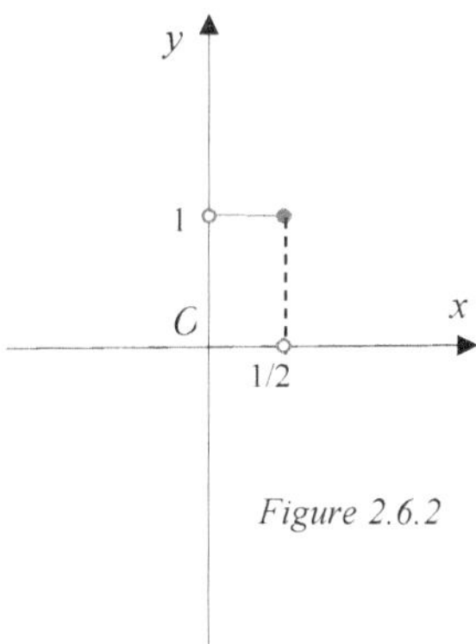

Figure 2.6.2

Graph the function $y_3 = \dfrac{x(1-2x)}{2x+1} \quad \forall x: x > 1/2$

DOMAIN

$$2x + 1 \neq 0 \;\Rightarrow\; x \neq -1/2 \;\Rightarrow\; D_3 = \,]1/2, +\infty[.$$

SIGN OF FUNCTION

$$y_3 \geq 0 \quad \Rightarrow \quad \frac{x(1-2x)}{2x+1} \geq 0 \quad \Rightarrow \quad \forall x \in \varnothing \cap D_3 = \varnothing$$

LIMITS AND ASYMPTOTES

$$\lim_{x \to +\infty} y = \lim_{x \to +\infty} \frac{x(1-2x)}{2x+1} = +\infty$$

$$m = \lim_{x \to +\infty} \frac{y}{x} = \lim_{x \to +\infty} \frac{x(1-2x)}{2x+1}\frac{1}{x} = -1, \quad n = \lim_{x \to +\infty} y - mx = \lim_{x \to +\infty} \frac{x(1-2x)}{2x+1} + x = 1 \Rightarrow$$

$$\Rightarrow y = -x + 1 \quad \text{oblique asymptote } (x \to +\infty).$$

$$\lim_{x \to \frac{1}{2}^+} y = \lim_{x \to \frac{1}{2}^+} \frac{x(1-2x)}{2x+1} = 0 \;\Rightarrow P(1/2,\,0) \text{ removable discontinuity}$$

<u>THE FIRST ORDER DERIVATIVE</u>

$$y' = \frac{-4x^2 - 4x + 1}{(2x+1)^2}$$

$$y' \geq 0 \quad \Rightarrow \quad \frac{-4x^2 - 4x + 1}{(2x+1)^2} \geq 0 \quad \Rightarrow -4x^2 - 4x + 1 \geq 0 \;\Rightarrow \forall x \in \varnothing \;\Rightarrow y \text{ is decreasing}$$

The graph of function y_3 is represented in Figure 2.6.3.

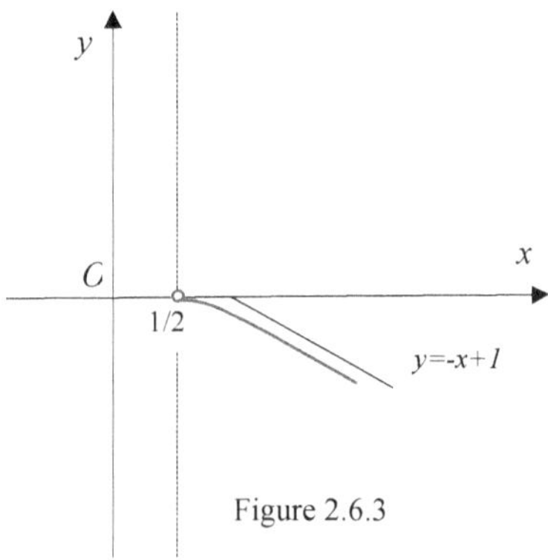

Figure 2.6.3

The graph of function $y = y_1 \cup y_2 \cup y_3$ is represented in Figure 2.6.

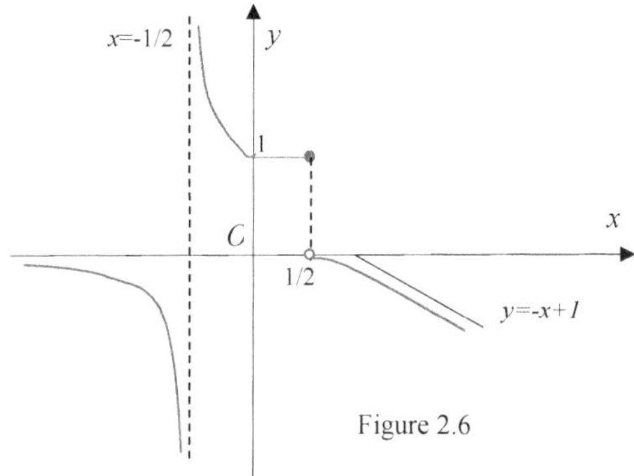

Figure 2.6

$x = 1/2$ jump discontinuity

N.7.- Graph the function $y = \dfrac{x^2 - 5x + 6}{x^2 - 4}$.

DOMAIN D

$$x^2 - 4 \neq 0 \ \Rightarrow x \neq \pm 2 \ \Rightarrow D = R - \{-2, +2\}$$

SIGN OF FUNCTION

$$y \geq 0 \ \Rightarrow \ \frac{x^2 - 5x + 6}{x^2 - 4} \geq 0 \ \Rightarrow \ \begin{cases} x^2 - 5x + 6 \geq 0 \\ x^2 - 4 > 0 \end{cases} \cup \begin{cases} x^2 - 5x + 6 \leq 0 \\ x^2 - 4 < 0 \end{cases} \Rightarrow$$

$$\Rightarrow x < -2 \vee x \geq 3$$

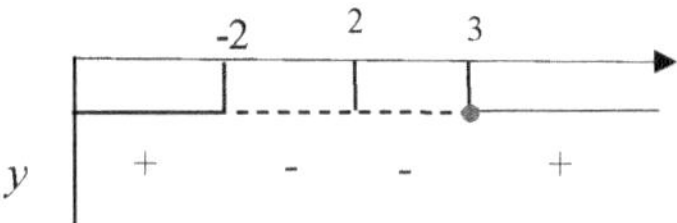

LIMITS AND ASYMPTOTES

$$\lim_{x \to 2^{\pm}} \frac{x^2 - 5x + 6}{x^2 - 4} = \left(\frac{0}{0}\right) = \lim_{x \to 2^{\pm}} \frac{(x-2)(x-3)}{(x-2)(x+2)} = \lim_{x \to 2^{\pm}} \frac{x-3}{x+2} = -\frac{1}{4} \ \Rightarrow$$

$$\Rightarrow P(2; -1/4) \text{ removable discontinuity.}$$

$$\lim_{x \to -2^+} \frac{x^2 - 5x + 6}{x^2 - 4} = -\infty, \qquad \lim_{x \to -2^-} \frac{x^2 - 5x + 6}{x^2 - 4} = +\infty \ \Rightarrow$$

$$\Rightarrow \ x = -2 \text{ vertical asymptote.}$$

$$\lim_{x \to \pm\infty} \frac{x^2 - 5x + 6}{x^2 - 4} = 1 \ \Rightarrow \ y = 1 \text{ horizontal asymptote.}$$

INTERSECTION OF THE GRAPH WITH THE COORDINATE AXES

$$\begin{cases} y = \dfrac{x^2 - 5x + 6}{x^2 - 4} \\ x = 0 \end{cases} \ \Rightarrow \ A(0, -3/2)$$

$$\begin{cases} y = \dfrac{x^2 - 5x + 6}{x^2 - 4} \\ y = 0 \end{cases} \Rightarrow \begin{cases} x^2 - 5x + 6 = 0 \\ y = 0 \end{cases} \Rightarrow \begin{cases} x = 3 \\ y = 0 \end{cases} \Rightarrow B(3,0)$$

THE FIRST ORDER DERIVATIVE

$$y' = \frac{(2x - 5)(x^2 - 4) - (x^2 - 5x + 6)2x}{(x^2 - 4)^2} = \frac{5}{(x + 2)^2}$$

$$y' \geq 0 \quad \Rightarrow \quad \frac{5}{(x + 2)^2} \geq 0 \quad \Rightarrow \quad \forall x \in D \quad \Rightarrow \quad y \text{ is decreasing}$$

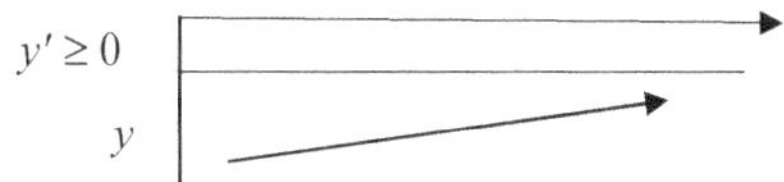

THE SECOND ORDER DERIVATIVE

$$y'' = \frac{-10(x + 2)}{(x + 2)^4}$$

$$y'' \geq 0 \quad \Rightarrow \quad \frac{-10(x + 2)}{(x + 2)^4} \geq 0 \quad \Rightarrow \quad x + 2 \leq 0 \quad \Rightarrow \quad x \leq -2 \quad \Rightarrow$$

$$\Rightarrow y \text{ is convex } \forall x \in \left] -\infty, \frac{1}{2} \right[$$

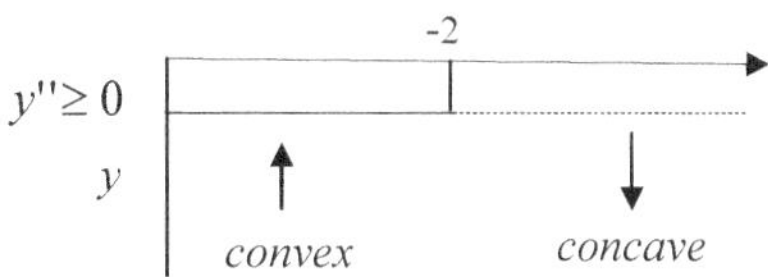

The graph of function is represented in .Figure 2.7.

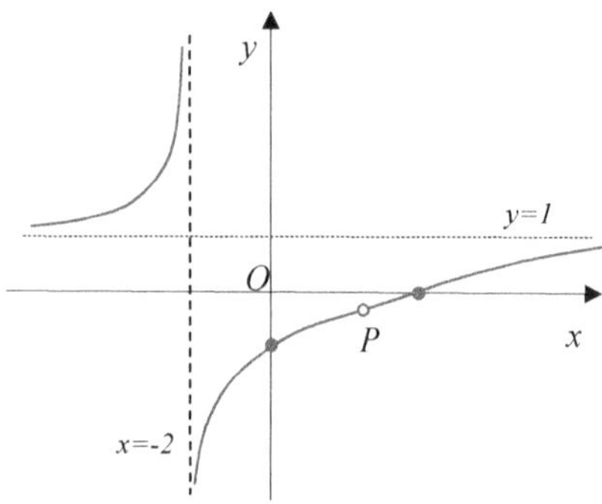

Figure 2.7

N.8.- Graph the function $y = \dfrac{x^3 + 1}{x^2 - 1}$.

DOMAIN *D*

$$x^2 - 1 \neq 0 \quad \Rightarrow \quad x \neq \pm 1 \Rightarrow D = R - \{ -1, +1 \}.$$

SIGN OF FUCTION

$$y \geq 0 \quad \Rightarrow \quad \frac{x^3 + 1}{x^2 - 1} \geq 0 \quad \Rightarrow \quad \frac{(x+1)(x^2 - x + 1)}{(x-1)(x+1)} \geq 0 \quad \Rightarrow$$

$$\Rightarrow \quad \frac{x^2 - x + 1}{x - 1} \geq 0 \quad \Rightarrow \quad x - 1 > 0 \quad \Rightarrow \quad x > 1$$

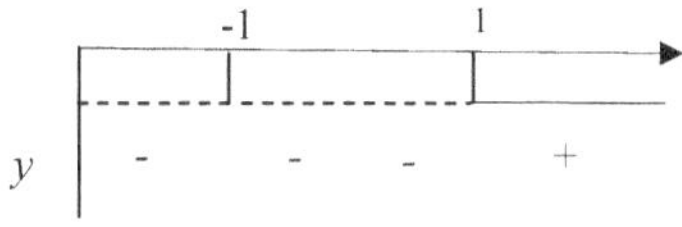

LIMITS AND ASYMPTOTES

$$\lim_{x \to 1^{\pm}} \frac{x^3 + 1}{x^2 - 1} = +\infty, \quad \Rightarrow x = 1 \text{ vertical asymptote}$$

$$\lim_{x \to -1^{\pm}} \frac{x^3 + 1}{x^2 - 1} = -\frac{3}{2}, \quad \Rightarrow \quad P(-1,-3/2) \text{ removable discontinuity}$$

$$\lim_{x \to \pm\infty} \frac{x^3 + 1}{x^2 - 1} = \pm\infty \quad \Rightarrow y = x \text{ oblique asymptote:}$$

$$m = \lim_{x \to \pm\infty} \frac{\frac{x^3 + 1}{x^2 - 1}}{x} = \lim_{x \to \pm\infty} \frac{x^3 + 1}{x^3 - x} = 1,$$

$$n = \lim_{x \to \pm\infty} \frac{x^3 + 1}{x^2 - 1} - x = \lim_{x \to \pm\infty} \frac{x^3 + 1 - x^3 + x}{x^2 - 1} = \lim_{x \to \pm\infty} \frac{1}{x - 1} = 0$$

THE FIRST ORDER DERIVATIVE

$$y' = \frac{3x(x^2-1)-(x^3+1)2x}{(x^2-1)^2} = \frac{x^4-3x^2-2x}{(x^2-1)^2} = \frac{x(x-2)(x+1)^2}{\left(x^2-1\right)^2}$$

$$y' \geq 0 \quad \Rightarrow \quad \frac{x(x-2)(x+1)^2}{\left(x^2-1\right)^2} \geq 0 \quad \Rightarrow \quad x(x-2) \geq 0 \Rightarrow x \leq 0 \ \lor \ x \geq 2$$

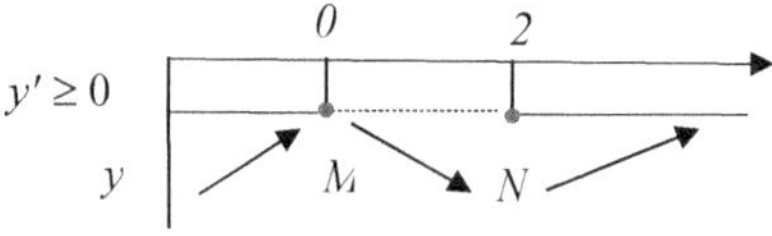

Local minimums and maximums : $M(0,-1)$, $N(2,3)$.

THE SECOND ORDER DERIVATIVE

$$y'' = \frac{(4x^3-6x-2)(x^2-1)^2-(x^4-3x^2-2x+1)2\left(x^2-1\right)2x}{(x^2-1)^4} = \frac{2(x-1)(x+1)^4}{\left(x^2-1\right)^4}$$

$$y'' \geq 0 \quad \Rightarrow \quad \frac{2(x-1)(x+1)^4}{\left(x^2-1\right)^4} \geq 0 \quad \Rightarrow \quad \frac{2}{(x-1)^3} \geq 0 \quad \Rightarrow \quad x-1>0 \quad \Rightarrow \quad x>1$$

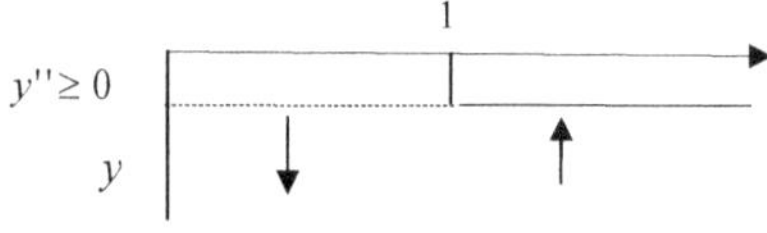

The graph of function is represented in .Figure 2.8.

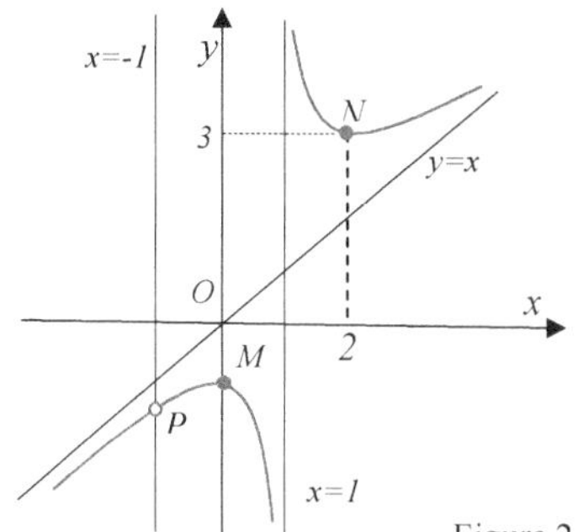

Figure 2.8

Exercises .- Graph the functions.

a) $y = \dfrac{x^3 - 3x^2 + 4}{x^2}$;

b) $y = \dfrac{4x^2 + 1}{3x}$;

c) $y = \left(\dfrac{1}{x} - 1\right)^2$;

d) $y = \dfrac{3x^2 - 2\sqrt{3}x - 3}{x^2 + 1}$;

e) $y = \begin{cases} -\dfrac{(2x-1)^3}{6x^2} & \forall x : x < \dfrac{1}{2} \\[2mm] \dfrac{(2x-1)^3}{6x^2} & \forall x : x \ge \dfrac{1}{2} \end{cases}$;

f) $y = \dfrac{2(x^2 - 1)}{5x^2 + 4x}$;

g) $y = \dfrac{-2x^2 + 3}{x^2 - 2x + 2}$;

h) $y = \dfrac{2x^4 + x^3}{(2x+1)(-x^2 + 1)}$;

i) $y = \begin{cases} \dfrac{x^3}{1 - x^2} & \forall x : x \ge 0 \\[2mm] \dfrac{x^3 - 1}{x^2} & \forall x : x < 0 \end{cases}$;

l) $y = \begin{cases} \dfrac{4x}{1 + x^2} & \forall x : x \le 1 \\[2mm] \dfrac{x + 1}{x^2 - 2x + 1} & \forall x : x > 1 \end{cases}$;

m) $y = \dfrac{3x^4 + 2x^2 + 7}{2x^4 + x^3 + 4x}$;

n) $y = \dfrac{1}{x + 2} - \dfrac{1}{x^2 - 4}$;

o) $y = \dfrac{x^3 - x + 5}{x^4 + 2x^2 - 3}$;

p) $y = \dfrac{x^3 + \dfrac{2}{x^2}}{2x^3 - \dfrac{1}{x^2}}$;

q) $y = \dfrac{x\sqrt{2}}{x + 1} - \dfrac{x\sqrt{3} - x + 5}{x^2 - 1}$.

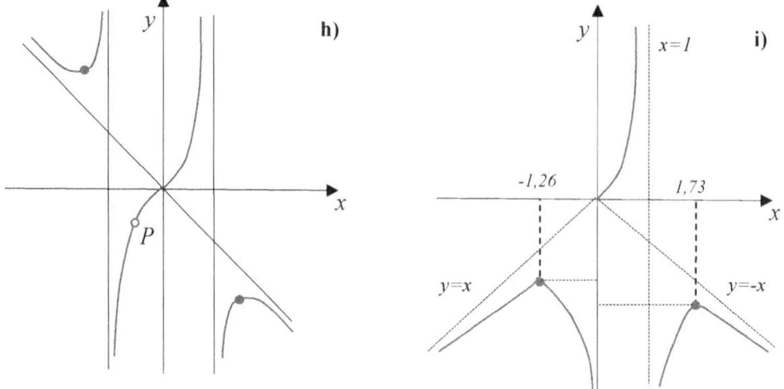

P(-1/2,-1/3) is a removable discontinuity.

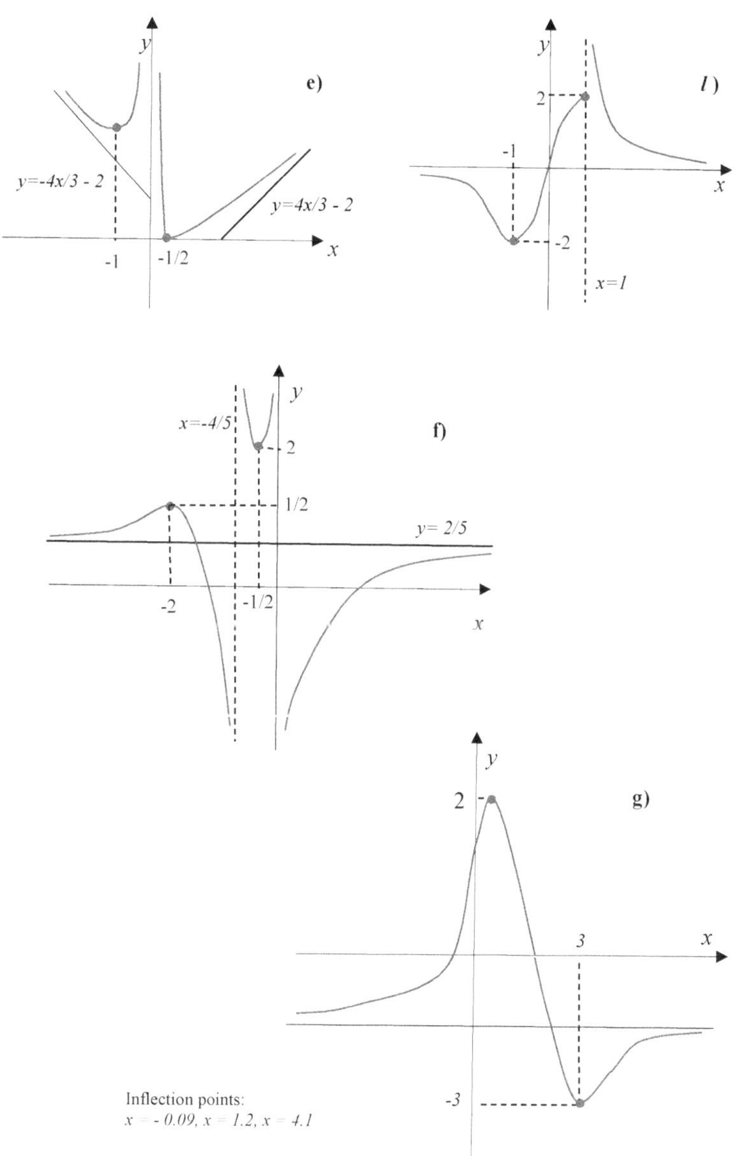
e)
y=-4x/3 - 2
y=4x/3 - 2
-1
-1/2
x
y

l)
2
-1
-2
x=1
x
y

x=-4/5
2
1/2
y= 2/5
-2
-1/2
x
y
f)

g)
2
3
-3
x
y

Inflection points:
x = - 0.09, x = 1.2, x = 4.1

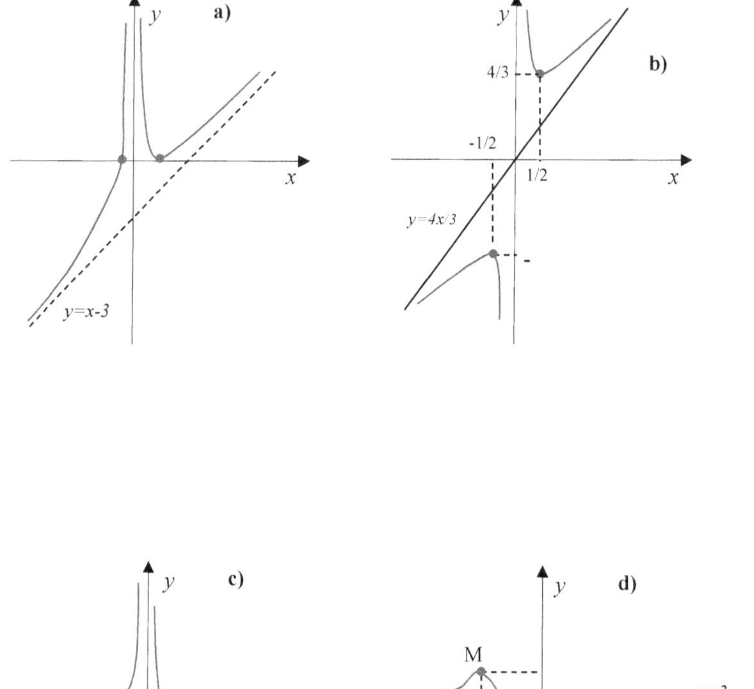

$$N\left(2+\sqrt{3},\,-2\sqrt{3}\right), \quad M\left(-2-\sqrt{3},\,2\sqrt{3}\right)$$

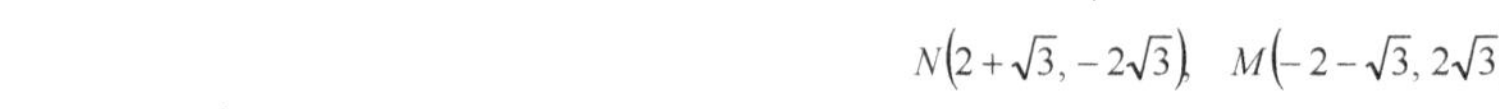

www.matematicus.com

3. Graphs of irrational functions.

N.1. - Graph the function $y = \sqrt{\dfrac{2-x}{x}}$.

DOMAIN D

$$\frac{2-x}{x} \geq 0 \quad \Rightarrow \quad \begin{cases} 2-x \geq 0 \\ x > 0 \end{cases} \cup \begin{cases} 2-x \leq 0 \\ x < 0 \end{cases} \Rightarrow \quad 0 < x \leq 2 \quad \Rightarrow \quad D = \left]0,2\right]$$

SIGN OF FUCTION

$y \geq 0 \qquad \forall x \in \left]0;2\right]$

$y = 0 \Rightarrow x = 2$

LIMITS AND ASYMPTOTES

$$\lim_{x \to 0^+} \sqrt{\frac{2-x}{x}} = \sqrt{\frac{2-0}{0}} = \frac{2}{0^+} = +\infty \quad \Rightarrow \quad x = 0 \text{ vertical asymptote}$$

THE FIRST ORDER DERIVATIVE

$$y' = \frac{1}{2}\left(\frac{2-x}{x}\right)^{-\frac{1}{2}} D\left(\frac{2-x}{x}\right) = \frac{-1}{\sqrt{x^3(2-x)}} \quad , \quad x \neq 2 .$$

$y' < 0 \quad \forall x \in \left]0,2\right[\quad \Rightarrow \quad y \text{ is dcreasing}$

$$\lim_{x \to 2^-} \frac{-1}{\sqrt{x^3(2-x)}} = -\infty$$

THE SECOND ORDER DERIVATIVE

$$y'' = \frac{0 - \left[-1 \cdot D\sqrt{x^3(2-x)}\right]}{\left(\sqrt{x^3(2-x)}\right)^2} = \frac{\dfrac{1}{2}\left[x^3(2-x)\right]^{-\frac{1}{2}} \cdot \left[3x^2(2-x) + x^3(-1)\right]}{x^3(2-x)} =$$

$$= \frac{6x^2 - 4x^3}{2\left[x^3(2-x)\right]\sqrt{x^3(2-x)}} = \frac{2x^2(3-2x)}{2\left[x^3(2-x)\right]\sqrt{x^3(2-x)}} = \frac{3-2x}{x(2-x)\sqrt{x^3(2-x)}} = \frac{3-2x}{\sqrt{x^5(2-x)^3}},$$

$$y'' \geq 0 \quad \Rightarrow \quad \frac{3-2x}{\sqrt{x^5(2-x)^3}} \geq 0 \quad \Rightarrow \quad 3-2x \geq 0 \quad \Rightarrow \quad x \leq \frac{3}{2}$$

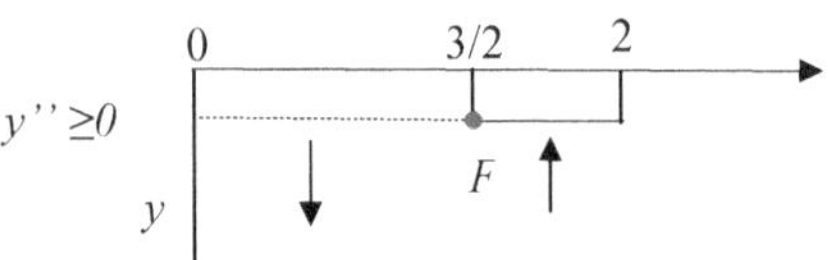

Inflection point: $F\left(\dfrac{3}{2}; \dfrac{\sqrt{3}}{3}\right)$

$$y\left(\frac{3}{2}\right) = \sqrt{\frac{2-\frac{3}{2}}{\frac{3}{2}}} = \sqrt{\frac{1}{2} \cdot \frac{2}{3}} = \sqrt{\frac{1}{3}} = \frac{\sqrt{1}}{\sqrt{3}} = \frac{1}{\sqrt{3}} = \frac{\sqrt{3}}{3} \quad \Rightarrow \quad F\left(\frac{3}{2}; \frac{\sqrt{3}}{3}\right)$$

The graph of function is represented in Figure 3.1

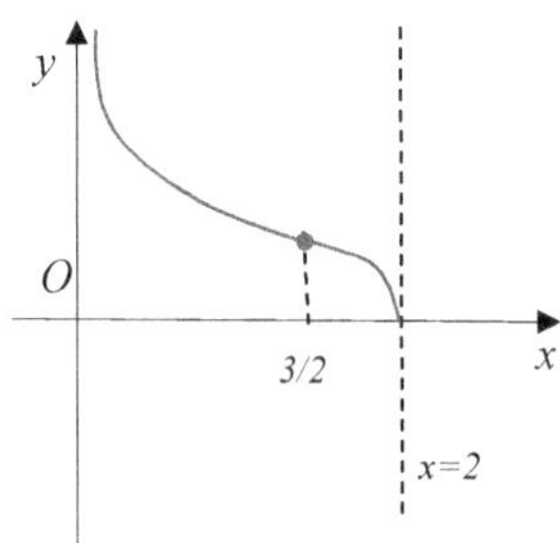

Figure 3.1

N.2.- Graph the function $y = \dfrac{x}{\sqrt{x-1}}$.

DOMAIN *D*

$$x - 1 > 0 \quad \Rightarrow \quad x > 1 \quad \Rightarrow \quad D = \,]1; +\infty\,[$$

SIGN OF FUCTION

$$y > 0 \quad \forall x \in D$$

LIMITS AND ASYMPTOTES

$$\lim_{x \to 1^+} \frac{x}{\sqrt{x-1}} = \frac{1}{0^+} = +\infty \quad \Rightarrow \quad x = 1 \quad \text{vertical asymptote}$$

$$\lim_{x \to +\infty} \frac{x}{\sqrt{x-1}} = \lim_{x \to +\infty} \sqrt{\frac{x^2}{x-1}} = \sqrt{\lim_{x \to +\infty} \frac{x^2}{x-1}} = +\infty \,,$$

$$m = \lim_{x \to +\infty} \frac{x}{\sqrt{x-1}} \cdot \frac{1}{x} = \lim_{x \to +\infty} \frac{1}{\sqrt{x-1}} = \frac{1}{+\infty} = 0,$$

THE FIRST ORDER DERIVATIVE

$$y' = \frac{\sqrt{x-1} - x\dfrac{1}{2}(x-1)^{\frac{-1}{2}}}{\sqrt{(x-1)^2}} = \frac{\sqrt{x-1} - \dfrac{x}{2\sqrt{x-1}}}{\sqrt{(x-1)^2}} = \frac{2(x-1) - x}{2\sqrt{(x-1)^3}} = \frac{1}{2}\frac{x-2}{\sqrt{(x-1)^3}}$$

$$y' \geq 0 \quad \Rightarrow \quad \frac{x-2}{2\sqrt{(x-1)^3}} \geq 0 \quad \Rightarrow \quad x - 2 \geq 0 \quad \Rightarrow \quad x \geq 2$$

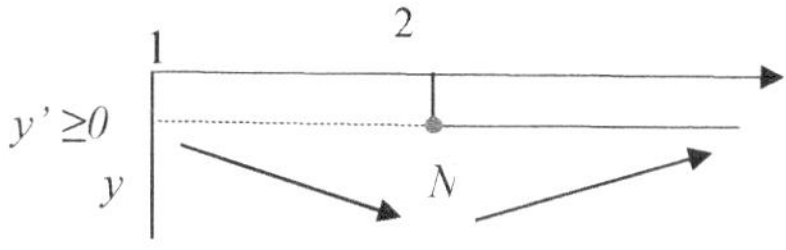

Local minimum: $N(2,2)$

$$y(2) = \frac{2}{\sqrt{2-1}} = 2 \quad \Rightarrow \quad N(2,2).$$

THE SECOND ORDER DERIVATIVE

$$y'' = \frac{2\sqrt{(x-1)^3} - (x-2)\cdot 2\left[\frac{3}{2}(x-1)^{\frac{1}{2}}\right]}{4\sqrt{(x-1)^6}} = \frac{2\sqrt{(x-1)^3} - 3(x-2)\sqrt{x-1}}{4\sqrt{(x-1)^6}} =$$

$$= \frac{2(x-1)\sqrt{x-1} - 3(x-2)\sqrt{x-1}}{4\sqrt{(x-1)^6}} = \frac{\sqrt{x-1}\left[2(x-1) - 3(x-2)\right]}{4\sqrt{(x-1)^6}} = \frac{4-x}{4\sqrt{(x-1)^5}}$$

$$y'' \geq 0 \quad \Rightarrow \quad \frac{4-x}{4\sqrt{(x-1)^5}} \geq 0 \quad \Rightarrow \quad 4-x \geq 0 \quad \Rightarrow \quad x \leq 4$$

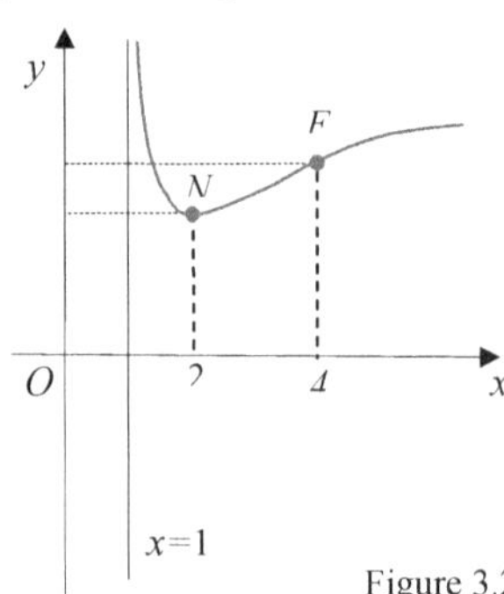

$$y(4) = \frac{4}{\sqrt{4-1}} = \frac{4}{\sqrt{3}} = \frac{4\sqrt{3}}{3} \quad \Rightarrow \quad F\left(4, \frac{4\sqrt{3}}{3}\right)$$

The graph of function is represented in Figure 3.2.

Figure 3.2

N.3.- Graph the function $y = \sqrt{\dfrac{x^2 - 2x}{x^2 - 1}}$.

DOMAIN D

$$\frac{x^2 - 2x}{x^2 - 1} \geq 0 \quad \Rightarrow \quad \begin{cases} x^2 - 2x \geq 0 \\ x^2 - 1 > 0 \end{cases} \cup \begin{cases} x^2 - 2x \leq 0 \\ x^2 - 1 < 0 \end{cases} \quad \Rightarrow \quad x < -1, \; 0 \leq x < 1, \; x \geq 2 \quad \Rightarrow$$

$$\Rightarrow \; D = \;]\text{-}\infty, \text{-}1[\; \cup \; [0,1[\; \cup \; [2, +\infty[.$$

SIGN OF FUCTION

$$y \geq 0 \quad \forall x \in D$$

$$x = 0 \;\Rightarrow\; y = 0$$
$$x = 2 \;\Rightarrow\; y = 0.$$

LIMITS AND ASYMPTOTES

$$\lim_{x \to -1^-} \sqrt{\frac{x^2 - 2x}{x^2 - 1}} = \sqrt{\frac{1 - 2(-1)}{1 - 1}} = \sqrt{\frac{3}{0^+}} = +\infty \quad \Rightarrow \quad x = -1 \;\; \text{vertical asymptote}$$

$$\lim_{x \to 1^-} \sqrt{\frac{x^2 - 2x}{x^2 - 1}} = \sqrt{\frac{1 - 2}{0^-}} = \sqrt{\frac{-1}{0^-}} = +\infty \quad \Rightarrow \quad x = 1 \;\; \text{vertical asymptote}$$

$$\lim_{x \to \pm\infty} \sqrt{\frac{x^2 - 2x}{x^2 - 1}} = \sqrt{\lim_{x \to \pm\infty} \frac{x^2 - 2x}{x^2 - 1}} = \sqrt{1} = 1 \quad \Rightarrow \quad y = 1 \;\; \text{horizontal asymptote}$$

THE FIRST ORDER DERIVATIVE

$$y' = \frac{1}{2}\left(\frac{x - 2x}{x - 1}\right)^{-\frac{1}{2}} \frac{(2x - 2)(x^2 - 1) - 2x(x^2 - 2x)}{\left(x^2 - 1\right)^2} = \frac{x^2 - x + 1}{\left(x^2 - 1\right)^2 \sqrt{\dfrac{x^2 - 2x}{x^2 - 1}}}$$

$$y' > 0 \quad \forall x \in D - \{0, 2\} \quad \Rightarrow \quad y \text{ is increasing}$$

$$\lim_{x \to 0^+} y' = +\infty, \qquad \lim_{x \to 2^+} y' = +\infty$$

The graph of function is represented in Figure 3.3

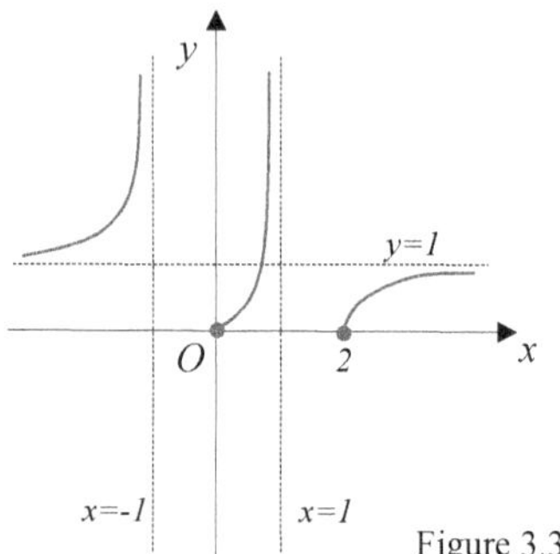

Figure 3.3

www.matematicus.com

N.4.- Graph the function $y = \sqrt[5]{x^2(5-x)^3}$.

DOMAIN D

$$D = R$$

SIGN OF FUCTION

$$y \geq 0 \quad \Rightarrow \quad \sqrt[5]{x^2(5-x)^3} \geq 0 \quad \Rightarrow \quad x^2(5-x)^3 \geq 0 \quad \Rightarrow \quad 5-x \geq 0 \cup x = 0 \quad \Rightarrow x \leq 5 \cup x = 0$$

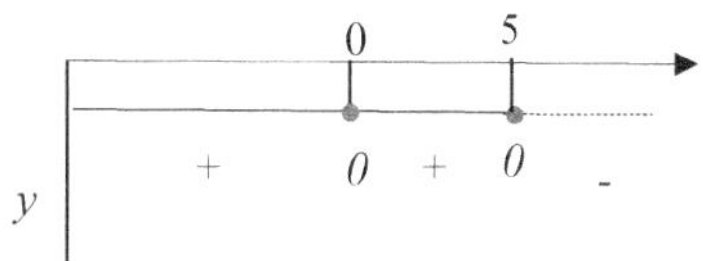

$$x = 0 \quad \Rightarrow \quad y = 0; \quad x = 5 \quad \Rightarrow \quad y = 0.$$

LIMITI E ASINITOTI

$$\lim_{x \to +\infty} \sqrt[5]{x^2(5-x)^3} = \sqrt[5]{(+\infty)^2(-\infty)^3} = \sqrt[5]{-\infty} = -\infty$$

$$\lim_{x \to -\infty} \sqrt[5]{x^2(5-x)^3} = \sqrt[5]{(-\infty)^2(+\infty)^3} = \sqrt[5]{+\infty} = +\infty;$$

$$m = \lim_{x \to \pm\infty} \frac{\sqrt[5]{x^2(5-x)^3}}{x} = \lim_{x \to \pm\infty} \sqrt[5]{\frac{x^2(5-x)^3}{x^5}} = \lim_{x \to \pm\infty} \sqrt[5]{\left(\frac{5-x}{x}\right)^3} = \lim_{x \to \pm\infty} \sqrt[5]{-1} = -1,$$

$$n = \lim_{x \to \pm\infty} \sqrt[5]{x^2(5-x)^3} + x = \lim_{x \to \pm\infty} x\left(\frac{\sqrt[5]{x^2(5-x)^3}}{x} + 1\right) = (+\infty) \cdot 0 \overset{DH}{=} \lim_{x \to \pm\infty} \left(\frac{\sqrt[5]{x^2(5-x)^3} + 1}{\frac{1}{x}}\right) =$$

$$= \lim_{x \to \pm\infty} \frac{\frac{3}{5}\left(\frac{5-x}{x}\right)^{-\frac{2}{5}} \cdot \frac{-x-5+x}{x^2}}{-\frac{1}{x^2}} = \lim_{x \to \pm\infty} \frac{3}{\sqrt[5]{\left(\frac{5-x}{x}\right)^2}} = \frac{3}{\lim_{x \to \pm\infty} \sqrt[5]{\left(\frac{5-x}{x}\right)^2}} = \frac{3}{1} = 3 \quad \Rightarrow$$

$$\Rightarrow y = -x + 3 \text{ oblique asymptote.}$$

THE FIRST ORDER DERIVATIVE

$$y' = \frac{1}{5}\left[x^2(5-x)^3\right]^{\frac{1}{5}-1} \cdot \left[2x(5-x)^3 + x^2[3(5-x)^2(-1)]\right] = \frac{2x(5-x)^3 - 3x^2(5-x)^2}{5 \cdot \sqrt[5]{[x^2(5-x)^3]^4}} =$$

$$= \frac{(5-x)^2(2x-x^2)}{5 \cdot \sqrt[5]{[x^2(5-x)^3]^4}}$$

$$y' \geq 0 \Rightarrow \frac{(5-x)^2(2x-x^2)}{5 \cdot \sqrt[5]{[x^2(5-x)^3]^4}} \geq 0 \Rightarrow (5-x)^2(2x-x^2) \geq 0 \Rightarrow 0 \leq x \leq 2 \ \lor \ x = 5$$

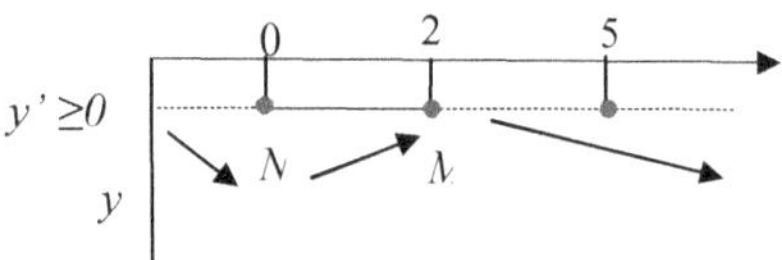

Local minimums and maximums: $M\left(2, \sqrt[5]{108}\right)$, $N(0,0)$

$$y(2) = \sqrt[5]{(2)^2(5-2)^3} = \sqrt[5]{108} \approx 2{,}55 \ \Rightarrow \ M\left(2, \sqrt[5]{108}\right); \quad y(0) = 0 \Rightarrow \ N(0,0)$$

Other points of function: $A\left(4, \sqrt[5]{16} \approx 1{,}74\right), B\left(1, \sqrt[5]{64} \approx 2{,}29\right)$

The graph of function is represented in Figure 3.4

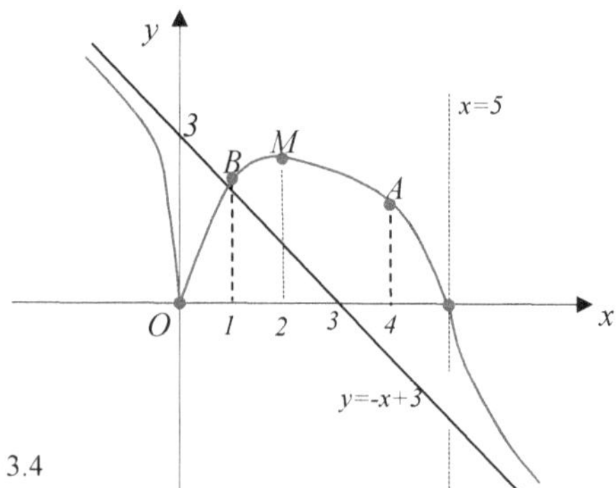

Figure 3.4

N.5.- Graph the function $y = \begin{cases} \sqrt{x} & \forall x : x \geq 0 \\ -2x^2 - 4x + 1 & \forall x : x < 0 \end{cases}$.

$$y = y_1 \cup y_2$$

$$y_1 = \sqrt{x} \qquad \forall x : x \geq 0$$

$$y_2 = -2x^2 - 4x + 1 \qquad \forall x : x < 0$$

The graph of function y_1 is represented in Figure 3.5.1, the graph of function y_2 is represented in Figure 3.5.2 and the graph of function $y = y_1 \cup y_2$ is represented in Figure 3.5.

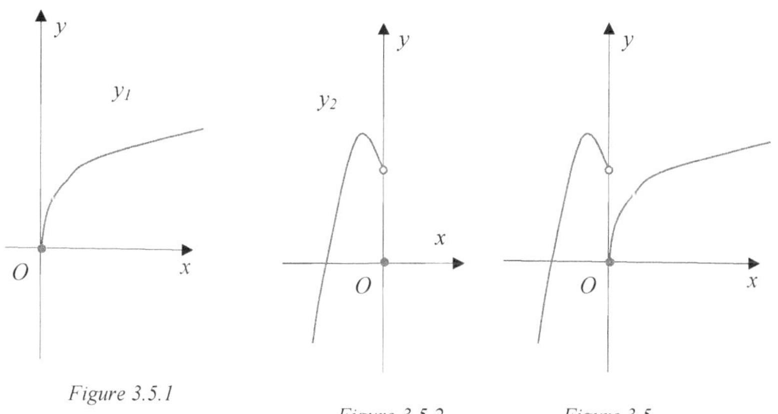

Figure 3.5.1

Figure 3.5.2

Figure 3.5

www.matematicus.com

N.6.- Graph the function $y = \begin{cases} \sqrt{\dfrac{x+1}{3x-1}} & \forall x : x \le -1 \\[2mm] x\sqrt{1-x^2} & \forall x : -1 < x \le 1 \end{cases}$

$$y = y_1 \cup y_2$$

$$y_1 = \sqrt{\frac{x+1}{3x-1}} \qquad \forall x : \quad x \le -1,$$

$$y_2 = x\sqrt{1-x^2} \qquad \forall x : \quad -1 < x \le 1$$

Graph the function $y_1 = \sqrt{\dfrac{x+1}{3x-1}} \quad \forall x : \ x \le -1$

DOMAIN D_1

$$\frac{x+1}{3x-1} \ge 0 \Rightarrow \begin{cases} x+1 \ge 0 \\ 3x-1 > 0 \end{cases} \cup \begin{cases} x+1 \le 0 \\ 3x-1 < 0 \end{cases} \Rightarrow D_1 =]-\infty, -1]$$

SIGN OF FUCTION

$$y_1 \ge 0 \quad \Rightarrow \quad \sqrt{\frac{x+1}{3x-1}} \ge 0 \Rightarrow \forall x \in D_1$$

LIMITS AND ASYMPTOTES

$$\lim_{x\to-\infty} y_1 = \lim_{x\to-\infty} \sqrt{\frac{x+1}{3x-1}} = \sqrt{\lim_{x\to-\infty} \frac{x+1}{3x-1}} = \sqrt{1/3} \quad \Rightarrow \quad y = \frac{\sqrt{3}}{3} \quad \text{horizontal asymptote}$$

THE FIRST ORDER DERIVATIVE

$$y'_1 = \frac{1}{2\sqrt{\dfrac{x+1}{3x-1}}} \frac{3x-1-(x+1)3}{(3x-1)^2} = \frac{-2}{(3x-1)^2 \sqrt{\dfrac{x+1}{3x-1}}}, \quad x \ne -1$$

$$y' \ge 0 \Rightarrow \frac{-2}{(3x-1)^2 \sqrt{\dfrac{x+1}{3x-1}}} \ge 0 \quad \Rightarrow \quad \forall x \in \varnothing$$

The graph of function y_1 is represented in Figure 3.6.1.

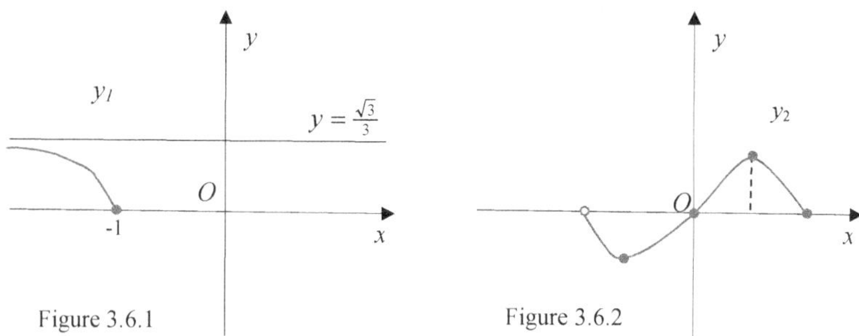

Figure 3.6.1   Figure 3.6.2

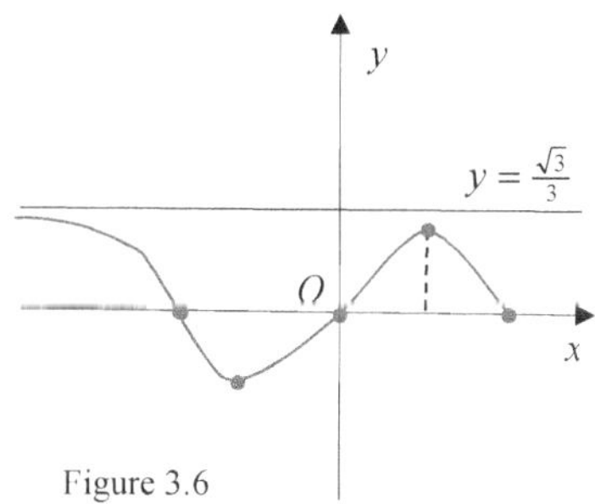

Figure 3.6

Graph the function $y_2 = x\sqrt{1-x^2}$ $\;\forall x:\; -1 < x \le 1$

DOMAIN D_2

$$1 - x^2 \ge 0 \quad \Rightarrow \; -1 < x \le 1 \Rightarrow D_2 = \,]\text{-}1,1].$$

SIGN OF FUCTION

$$y_2 \ge 0 \quad \Rightarrow \quad x\sqrt{1-x^2} \ge 0 \quad \Rightarrow \quad 0 \le x \le 1.$$

LIMITS AND ASYMPTOTES

$$\lim_{x \to -1} y_2 = \lim_{x \to -\infty} x\sqrt{1-x^2} = 0 \quad \Rightarrow \quad P(-1,\,0) \;\; \text{removable discontinuity}$$

THE FIRST ORDER DERIVATIVE

$$y'_2 = \sqrt{1-x^2} + \frac{x}{2\sqrt{1-x^2}}(-2x) = \frac{2(1-x^2)-2x^2}{2\sqrt{1-x^2}} = \frac{-2x^2+1}{\sqrt{1-x^2}}, \qquad x \neq 1$$

$$y'_2 \geq 0 \Rightarrow \frac{-2x^2+1}{\sqrt{1-x^2}} \geq 0 \Rightarrow -2x^2+1 \geq 0 \Rightarrow -\frac{\sqrt{2}}{2} \leq x \leq \frac{\sqrt{2}}{2}$$

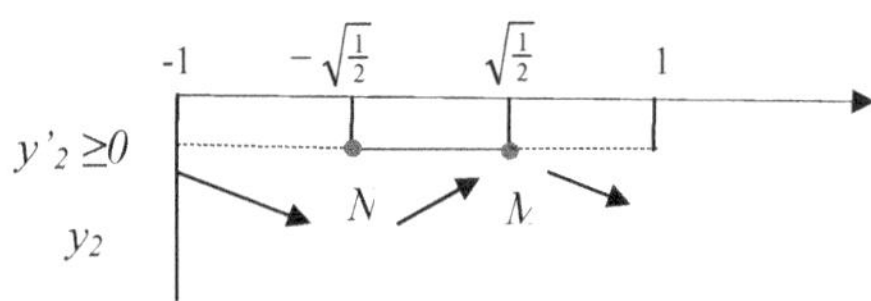

Local minimums and maximums: $M\left(\frac{\sqrt{2}}{2}; \frac{1}{2}\right), N\left(-\frac{\sqrt{2}}{2}; -\frac{1}{2}\right)$

$$\lim_{x\to 1^-} y' = -\infty$$

THE SECOND ORDER DERIVATIVE

$$y''_2 = \frac{-4x\sqrt{1-x^2} - (-2x^2+1)\dfrac{1}{2\sqrt{1-x^2}}(-2x)}{\left(\sqrt{1-x^2}\right)^2} = \frac{-8x(1-x^2)+2x(-2x^2+1)}{2\sqrt{1-x^2}\left(\sqrt{1-x^2}\right)^2} =$$

$$= \frac{-8x+8x^3-4x^3+2x}{2\sqrt{1-x^2}\left(\sqrt{1-x^2}\right)^2} = \frac{4x^3-6x}{2\sqrt{1-x^2}\left(\sqrt{1-x^2}\right)^2} = \frac{x(2x^2-3)}{\sqrt{1-x^2}\left(\sqrt{1-x^2}\right)^2}$$

$$y''_2 \geq 0 \Rightarrow \frac{x(2x^2-3)}{\sqrt{1-x^2}\left(\sqrt{1-x^2}\right)^2} \geq 0 \Rightarrow x(2x^2-3) \geq 0 \Rightarrow x = 0, x = \pm\sqrt{\frac{3}{2}}$$

Inflection point: F(0,0)

The graph of function y_2 is represented in Figure 3.6.2, and . the graph of function $y = y_1 \cup y_2$ is represented in Figure 3.6.

N.7.- Graph the function $y = \sqrt{x - x^2} - x$.

DOMAIN D

$$x - x^2 \geq 0 \quad \Rightarrow \quad 0 \leq x \leq 1 \quad \Rightarrow \quad D = [0,1].$$

SIGN OF FUNCTION

$$y \geq 0 \quad \Rightarrow \quad \sqrt{x - x^2} \geq x \quad \Rightarrow \quad x - x^2 \geq x^2 \quad \Rightarrow \quad x - 2x^2 \geq 0 \quad \Rightarrow \quad 0 \leq x \leq 1/2$$

$O(0,0)$, $A(1/2,0)$.

THE FIRST ORDER DERIVATIVE

$$y' = \frac{1}{2\sqrt{x - x^2}} D(x - x^2) - 1 = \frac{1 - 2x}{2\sqrt{x - x^2}} - 1 = \frac{1 - 2x - 2\sqrt{x - x^2}}{2\sqrt{x - x^2}}$$

$$y' \geq 0 \Rightarrow \frac{1 - 2x - 2\sqrt{x - x^2}}{2\sqrt{x - x^2}} \geq 0 \Rightarrow 1 - 2x \geq 2\sqrt{x - x^2} \Rightarrow \begin{cases} 1 - 2x > 0 \\ x - x^2 \geq 0 \\ (1 - 2x)^2 < x - x^2 \end{cases} \Rightarrow$$

$$\Rightarrow 0 \leq x \leq \frac{1}{2} - \frac{\sqrt{2}}{4}$$

Local maximum: $M\left(\dfrac{1}{2} - \dfrac{\sqrt{2}}{4} ; \dfrac{\sqrt{2}}{2} - \dfrac{1}{2}\right)$.

The graph of function is represented in Figure 3.7

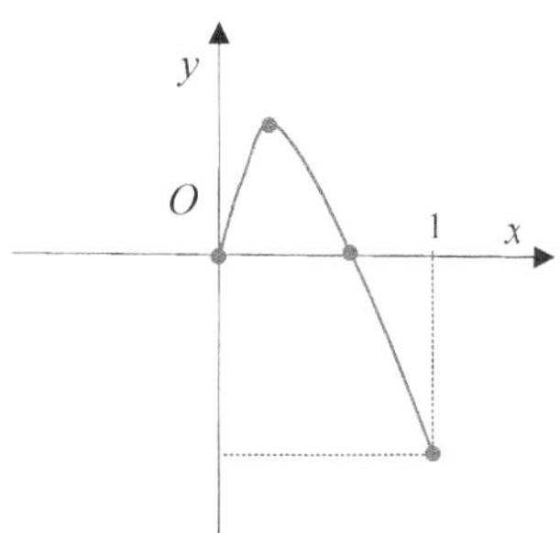

Figure 3.7

N.8.- Graph the function $y = \sqrt{-10x + x^2}$.

DOMAIN D

$$-10x + x^2 \geq 0 \quad \Rightarrow \quad x \leq 0, \; x \geq 10 \quad \Rightarrow \quad D =]-\infty, 0] \cup [10, +\infty[.$$

SIGN OF FUNCTION

$y > 0 \, \forall x \in D =]-\infty, 0] \cup [10, +\infty[$

$$x = 0 \quad \Rightarrow \quad y = 0; \quad x = 10 \quad \Rightarrow \quad y = 0.$$

$O(0,0), \; A(10,0)$

LIMITI ED ASINTOTI

$$\lim_{x \to \pm\infty} \sqrt{-10x + x^2} = +\infty$$

$$m = \lim_{x \to +\infty} \sqrt{-10x + x^2} \times \frac{1}{x} = \left(\frac{+\infty}{+\infty}\right) = \lim_{x \to +\infty} \sqrt{\frac{-10x + x^2}{x^2}} = \sqrt{\lim_{x \to +\infty} \frac{-10x + x^2}{x^2}} = 1$$

$$n = \lim_{x \to +\infty} \sqrt{-10x + x^2} - x = (+\infty - \infty) = \lim_{x \to +\infty} x\left(\sqrt{\frac{-10x + x^2}{x^2}} - 1\right) = (+\infty \cdot 0) =$$

$$= n = \lim_{x \to +\infty} \frac{\sqrt{\frac{-10x+x^2}{x^2}} - 1}{1/x} \overset{DH}{=} \lim_{x \to +\infty} \frac{-5}{\sqrt{1 - \frac{10}{x}}} = -5 \Rightarrow y = x\text{-}5. \quad \text{oblique asymptote}$$

$$y = -x + 5 \quad \text{oblique asymptote} \; (x \to -\infty)$$

THE FIRST ORDER DERIVATIVE.

$$y' = \frac{1}{2\sqrt{-10x + x^2}} D(-10x + x^2) = \frac{-5 + x}{\sqrt{-10x + x^2}}, \quad x \neq 0, x \neq 10$$

$$y' \geq 0 \Rightarrow \frac{-5 + x}{\sqrt{-10x + x^2}} \geq 0 \Rightarrow -5 + x \geq 0 \Rightarrow x \geq 5$$

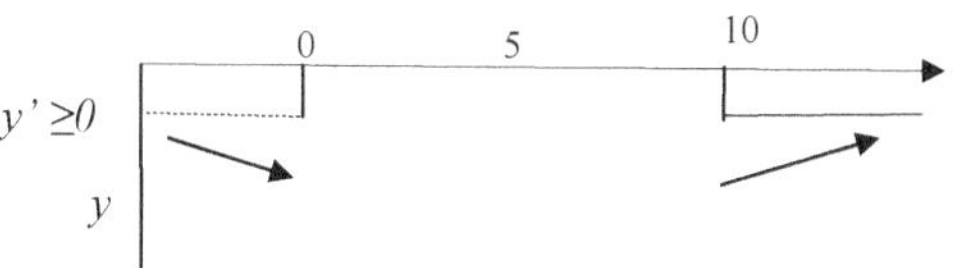

$$\lim_{x\to 0^-} y' = -\infty \,, \qquad \lim_{x\to 10^+} y' = +\infty$$

The graph of function is represented in Figure 3.8.

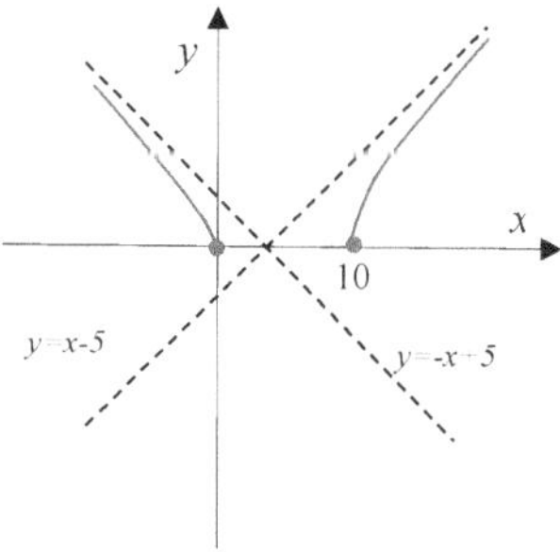

Figure 3.8

www.matematicus.com

N. 9. Graph the function $y = \dfrac{x^2 + 4x + 4}{\sqrt{x^2 + 3x - 6}}$.

DOMAIN D

$$x^2 + 3x - 6 > 0 \quad \Rightarrow \quad x < -\frac{3 + \sqrt{33}}{2} \approx -4{,}3, \quad x > \frac{-3 + \sqrt{33}}{2} \approx 1{,}3 \quad \Rightarrow$$

$$\Rightarrow \quad D = \left] -\infty; -\frac{3 + \sqrt{33}}{2} \right[\cup \left] \frac{-3 + \sqrt{33}}{2}, +\infty \right[.$$

SIGN OF FUNCTION

$y > 0 \quad \forall x \in D$

LIMITS AND ASYMPTOTES

$$\lim_{x \to -\frac{3+\sqrt{33}}{2}} y = \frac{m > 0}{0^+} = +\infty \quad \Rightarrow x = -\frac{3 + \sqrt{33}}{2} \quad \text{vertical asymptote}$$

$$\lim_{x \to \frac{3-\sqrt{33}}{2}} y = \frac{p > 0}{0^+} = +\infty \quad \Rightarrow x = \frac{3 - \sqrt{33}}{2} \quad \text{vertical asymptote}$$

$$\lim_{x \to \pm\infty} y = \lim_{x \to \pm\infty} \sqrt{\frac{(x + 2)^4}{x^2 + 3x - 6}} = \sqrt{+\infty} = +\infty$$

$$m = \lim_{x \to +\infty} y/x = \lim_{x \to +\infty} \frac{(x + 2)^2}{x\sqrt{x^2 + 3x - 6}} = \lim_{x \to +\infty} \sqrt{\frac{(x + 2)^2}{x^2(x^2 + 3x - 6)}} = \sqrt{1} = 1$$

$$n = \lim_{x \to +\infty} y - x = \lim_{x \to +\infty} x\left[\frac{(x^2 + 2)^2}{\sqrt{x^2 + 3x - 6}} - 1\right] = \lim_{x \to +\infty} \frac{\dfrac{(x + 2)^2}{\sqrt{x^2 + 3x - 6}} - 1}{1/x} \overset{DH}{=}$$

$$= \lim_{x \to +\infty} \frac{2(x+2)x\sqrt{x^2+3x-6} - (x+2)^2\left[\sqrt{x^2+3x-6} + x\dfrac{2x+3}{2\sqrt{x^2+3x-6}}\right]}{-1/x^2} = \quad \Rightarrow$$

$$= \lim_{x \to +\infty} \frac{(x+2)(5x^2+30x-24)}{2(x^2+3x-6)\sqrt{x^2+3x-6}} = \lim_{x \to +\infty} \frac{(x+2)(5x^2+30x-24)}{2x^3+6x^2+12x} \cdot \frac{1}{\sqrt{1+3/x-6/x^2}} = 5/2$$

$$\Rightarrow \quad y = x+5/2 \quad \text{oblique asymptote}$$

$y = -x-5/2$ is a oblique asymptote ($x \to -\infty$)

THE FIRST ORDER DERIVATIVE

$$y' = \frac{(2x+4)\sqrt{x^2+3x-6} - (x+2)^2\dfrac{2x+3}{2\sqrt{x^2+3x-6}}}{\sqrt{(x^2+3x-6)^2}} = \frac{(x+2)\left[2x+5x-30\right]}{2\sqrt{(x^2+3x-6)^3}}$$

$$y' \geq 0 \quad \Rightarrow \quad (x+2)\left[2x^2+5x-30)\right] \geq 0$$

$$x+2 \geq 0 \Rightarrow x \geq -2$$

$$2x^2+5x-6 \geq 0 \quad \Rightarrow \quad x \leq \frac{-5-\sqrt{265}}{4} \approx -5.32, \quad x \geq \frac{-5+\sqrt{265}}{4} \approx 2.82$$

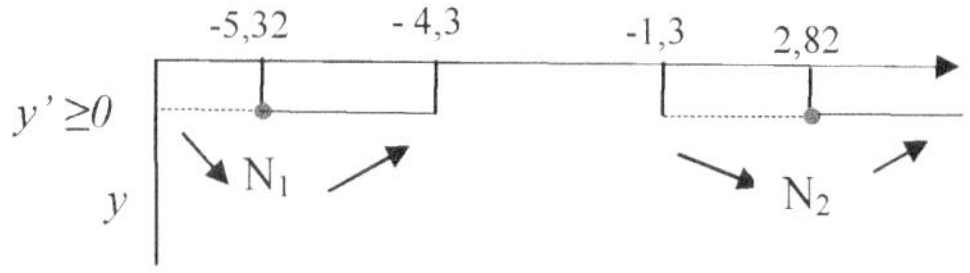

Local minimum $N_1(-5.32;4.38)$, $N_2 (2.82;7.19)$.

$$x = \frac{-5-\sqrt{265}}{4} \approx -5.32, \quad x = \frac{-5+\sqrt{265}}{4} \approx 2.82:$$

The graph of function is represented in Figure 3.9.

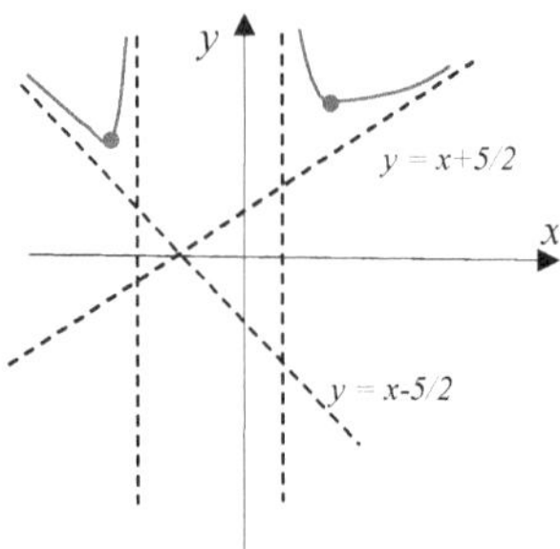

Figure 3.9

Exercises .- Graph the functions.

a) $y = \sqrt[3]{x(x^2 + 5x + 8)}$,   **b)** $y = \dfrac{4\sqrt{x+5}}{4\sqrt{x+5}}$,   **c)** $y = \sqrt{x^2 - 4x - 5}$,   **d)** $y = \dfrac{x^2 - 2x}{\sqrt{x^2 + 2x}}$,

e) $y = x - \sqrt{2x - 1}$,   **f)** $y = \dfrac{\sqrt{x^2 + x - 2}}{x - 1}$,   **g)** $y = (x + 3)\sqrt{\dfrac{x}{x+3}}$,   **h)** $y = \dfrac{x + 1}{\sqrt{2x^2 + x - 1}}$,

i) $y = \dfrac{x^2 + 4x + 4}{\sqrt{x^2 + 5x - 6}}$,   **l)** $y = \begin{cases} \dfrac{x-5}{2x-4} & \forall x : x \le 1/2 \\[2ex] x\sqrt{\dfrac{2x+1}{2x-1}} & \forall x : x > 1/2 \end{cases}$,   **m)** $y = x - 3 - \sqrt{x^2 - 9}$,

n) $y = \dfrac{x - 3}{x}\sqrt{\dfrac{2x - 1}{x - 3}}$,   **o)** $y = \dfrac{x(x^2 - 1)}{\sqrt{x}}$,   **p)** $y = -x + 4 - \sqrt[3]{x^2 - 3x + 13}$,

q) $y = \dfrac{x^2 + x - 6}{4\sqrt{2 - x}}$,   **r)** $y = \dfrac{\sqrt{4 - x^2}}{x^3}$,   **s)** $y = -2x - 2 - \sqrt[3]{x + 1}$,

t) $y = \begin{cases} x\sqrt{\dfrac{x-1}{x+1}} & \forall x : x > 1 \\[2ex] \sqrt[3]{-x^3 + 2x^2 - x} & \forall x : x \le 1 \end{cases}$,   **u)** $y = \sqrt{x^2 - 4x}$

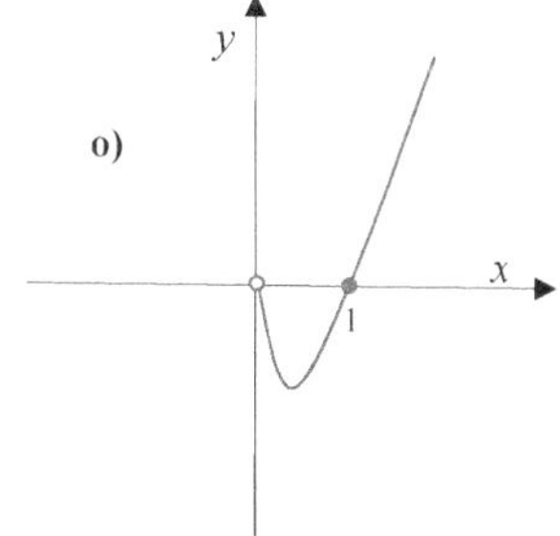

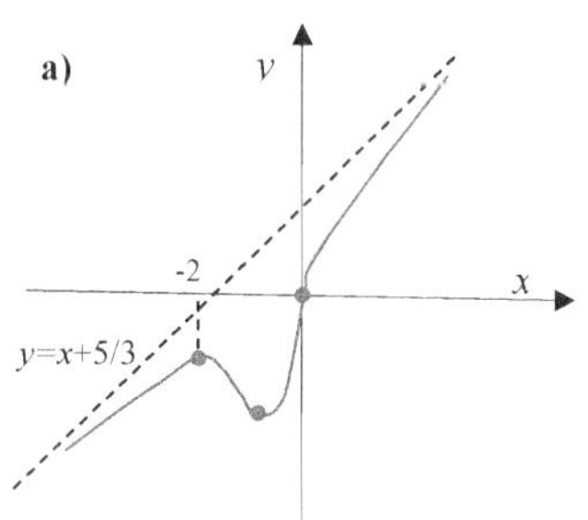

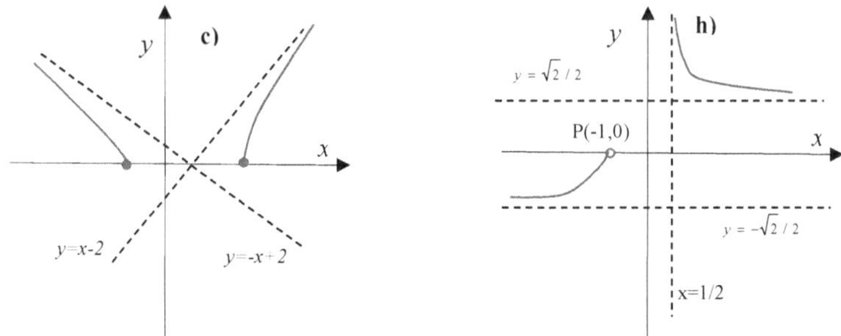
c)
y
x
y=x-2
y=-x+2
h)
y
x
y = √2 / 2
P(-1,0)
y = -√2 / 2
x=1/2

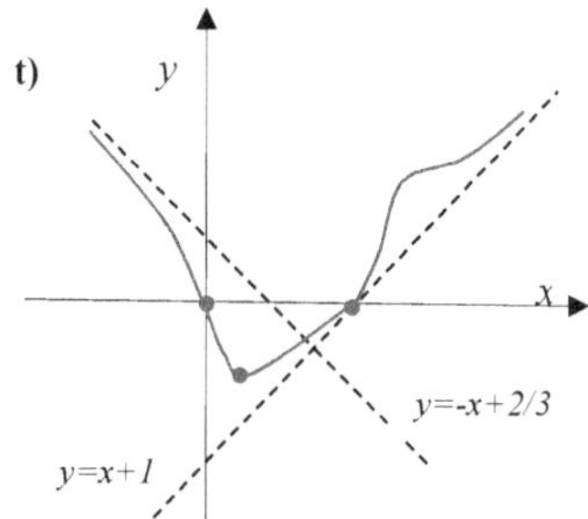
t)
y
x
y=-x+2/3
y=x+1

4. Graph of trigonometric functions.

N.1.- Graph the function $y = 2\sin x + \sin 2x$.

DOMAIN D

$$D = R$$

SYMMETRIE AND PERIODICITIES

$$y(x +2\pi) = 2\sin (x +2\pi) + \sin 2(x +2\pi) = 2\sin x + \sin 2x = y \;\Rightarrow$$

$$\Rightarrow y \text{ is a periodic function, period } p = 2\pi.$$

Graph the function $y = 2\sin x + \sin 2x \quad \forall x \in \left[0,\, 2\pi\right]$

SIGN OF FUCTION

$$y \geq 0 \;\Rightarrow\; 2\sin x + \sin 2x \geq 0 \;\Rightarrow\; 2\sin x + 2\sin x \cos x \geq 0 \;\Rightarrow$$

$$\Rightarrow 2ien x\,(\,1 + \cos x) \;\Rightarrow\; \begin{cases} \sin x \geq 0 \\ 1 + \cos x \geq 0 \end{cases} \cup \begin{cases} \sin x \leq 0 \\ 1 + \cos x \leq 0 \end{cases} \;\Rightarrow\; 0 \leq x \leq \pi \;\vee\; x = 2\pi$$

$$x = \pi \;\Rightarrow y = 0 \;\Rightarrow\; (\pi,0)$$

$$y(0) = 2\sin 0 + \sin(2 \cdot 0) = 0 \qquad\Rightarrow\qquad O(0;0).$$

$$y(2\pi) = 2\sin(2\pi) + \sin(2 \cdot 2\pi) = 0 \;\Rightarrow\; A(2\pi;0$$

THE FIRST ORDER DERIVATIVE

$$y' = 2\left(2\cos^2 x + \cos x - 1\right)$$

$$y' \geq 0 \;\Rightarrow\; 2\cos^2 x + \cos x - 1 \geq 0 \;\Rightarrow\; x = \pi,.0 \leq x \leq \frac{\pi}{3} \;\vee\; \frac{5}{3}\pi \leq x \leq 2\pi$$

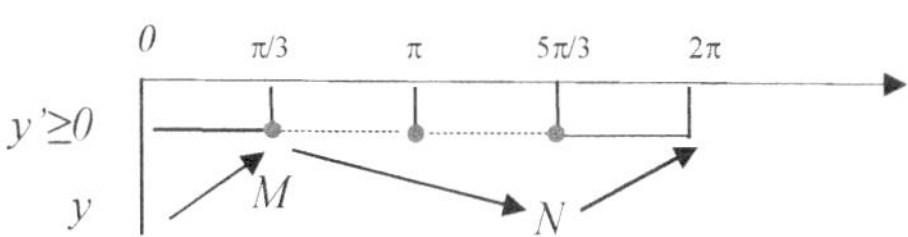

Local minimums and maximums: $M\left(\dfrac{\pi}{3};\dfrac{3}{2}\sqrt{3}\right)$, $N\left(\dfrac{5}{3}\pi;-\dfrac{3}{2}\sqrt{3}\right)$

$$y\left(\frac{\pi}{3}\right)=2\sin\left(\frac{\pi}{3}\right)+\sin 2\left(\frac{\pi}{3}\right)=2\frac{\sqrt{3}}{2}+2\sin\left(\frac{\pi}{3}\right)\cos\left(\frac{\pi}{3}\right)=\sqrt{3}+2\frac{\sqrt{3}}{2}\frac{1}{2}=\frac{3}{2}\sqrt{3}\,.$$

Inflection point: $x=\pi$

THE SECOND ORDER DERIVATIVE

$$y''=-2\left(2\sin 2x+\sin x\right)$$

$$y''\geq 0 \quad \Rightarrow \quad -2\left(2\sin 2x+\sin x\right)\geq 0 \quad \Rightarrow \quad 2\sin 2x+\sin x\leq 0 \quad \Rightarrow$$
$$\Rightarrow \quad \sin x\left(4\cos x+1\right)\leq 0 \quad \Rightarrow \alpha\leq x\leq\pi \ \vee\ 2\pi-\alpha\leq x\leq 2\pi,$$

$$\alpha=\arccos\left(-\frac{1}{4}\right)$$

Inflection points:

$$F_{1}\left(\alpha;\frac{3}{8}\sqrt{15}\right),\ F_{2}\left(\pi;0\right),\ F_{3}\left(2\pi-\alpha;-\frac{3}{8}\sqrt{15}\right)$$

The graph of function is represented in Figure 4.1.

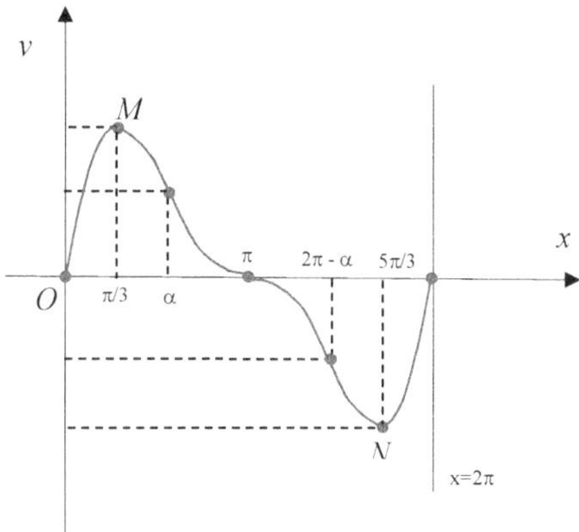

Figure 4.1

www.matematicus.com

N.2.- Graph the function $y = \sin^3 x + \cos^3 x$.

DOMAIN D

$$D = R$$

SYMMETRIE AND PERIODICITIES

$$y(x + 2\pi) = \sin^3 (x + 2\pi) + \cos^3 (x + 2\pi) = [\sin (x + 2\pi)]^3 + [\cos (x + 2\pi)]^3 =$$

$$= [\sin (x)]^3 + [\cos (x)]^3 = y \Rightarrow$$

$$\Rightarrow y \text{ is a periodic function, period } p = 2\pi.$$

Graph the function $y = \sin^3 x + \cos^3 x \quad \forall x \in [0, 2\pi]$

SIGN OF FUCTION

$$y \geq 0 \Rightarrow \sin^3 x + \cos^3 x \geq 0 \Rightarrow (\sin x + \cos x)(\sin^2 x - \sin x \cos x + \cos^2 x) \geq 0 \Rightarrow_2$$
$$\Rightarrow (\sin x + \cos x)(1 - \sin x \cos x) \geq 0 \Rightarrow \sin x + \cos x \geq 0$$

$$sin\, x + cos\, x \geq 0 \Rightarrow \quad cos\, x\, (tg\, x + 1) \geq 0 \Rightarrow$$

$$\Rightarrow \begin{cases} \cos x > 0 \\ \tan x + 1 \geq 0 \end{cases} \cup \begin{cases} \cos x < 0 \\ \tan x + 1 \leq 0 \end{cases} \Rightarrow$$

$$\Rightarrow 0 \leq x < \frac{3}{4}\pi \quad \vee \quad \frac{7}{4}\pi < x \leq 2\pi ,$$

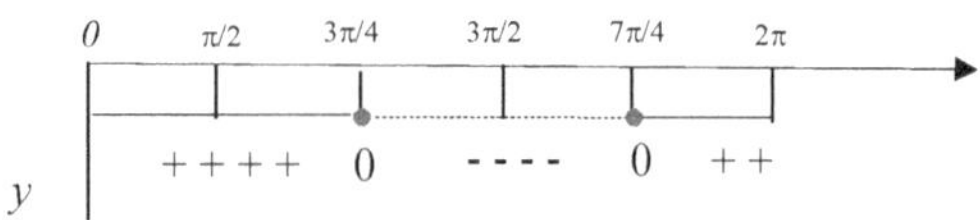

$$y(0) = y(2\pi) = 1, \Rightarrow A(0;1) , B(2\pi;1).$$

$$x = \frac{3}{4}\pi \Rightarrow y = 0 \quad x = \frac{7}{4}\pi \Rightarrow y = 0$$

---

[2] $1 - sen\, x\, cos\, x > 0 \ \forall x \in [0, 2\pi]$. www.matematicus.com

THE FIRST ORDER DERIVATIVE

$$y' = 3\sin x \cos x (\sin x - \cos x)$$

$$y' \geq 0 \Rightarrow 3\sin x \cos x (\sin x - \cos x) \geq 0 \Rightarrow \begin{cases} \sin x \cos x \geq 0 \\ \sin x - \cos x \geq 0 \end{cases} \cup \begin{cases} \sin x \cos x \leq 0 \\ \sin x - \cos x \leq 0 \end{cases} \Rightarrow$$

$$\Rightarrow \quad \frac{\pi}{4} \leq x \leq \frac{\pi}{2} \quad \vee \quad \pi \leq x \leq \frac{5}{4}\pi \quad \vee \quad \frac{3}{2}\pi \leq x \leq 2\pi$$

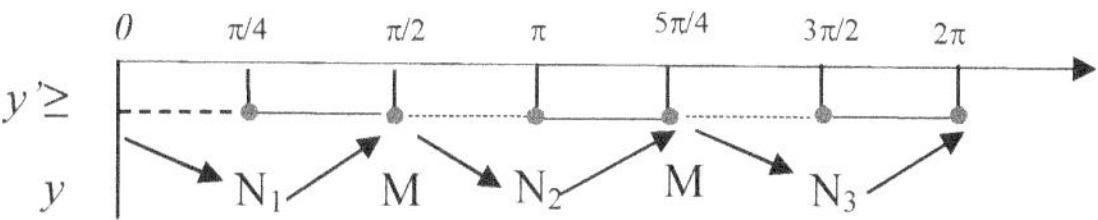

Local minimums and maximums:

$$M_1\left(\frac{\pi}{2};1\right), \quad M_2\left(\frac{5}{4}\pi;-\frac{\sqrt{2}}{2}\right), \quad N_1\left(\frac{\pi}{4};\frac{\sqrt{2}}{2}\right), \quad N_2\left(\pi;-1\right), \quad N_3\left(\frac{3}{2}\pi;-1\right)$$

$$y\left(\frac{\pi}{2}\right) = \sin^3\left(\frac{\pi}{2}\right) + \cos^3\left(\frac{\pi}{2}\right) - 1$$

$$y\left(\frac{5}{4}\pi\right) = \sin^3\left(\frac{5\pi}{4}\right) + \cos^3\left(\frac{5\pi}{4}\right) = -\frac{\sqrt{2}}{2}$$

$$y\left(\frac{\pi}{4}\right) = \sin^3\left(\frac{\pi}{4}\right) + \cos^3\left(\frac{\pi}{4}\right) = \left[\sin\left(\frac{\pi}{4}\right)\right]^3 + \left[\cos\left(\frac{\pi}{4}\right)\right]^3 = \left(\frac{\sqrt{2}}{2}\right)^3 + \left(\frac{\sqrt{2}}{2}\right)^3 = \frac{\sqrt{2}}{4} + \frac{\sqrt{2}}{4} = \frac{\sqrt{2}}{2}$$

$$y(\pi) = \sin^3(\pi) + \cos^3(\pi) = -1, \qquad y\left(\frac{3}{2}\pi\right) = \sin^3\left(\frac{3\pi}{2}\right) + \cos^3\left(\frac{3\pi}{2}\right) = -1$$

THE SECOND ORDER DERIVATIVE

$$y'' = 3(\sin x + \cos x)(3\sin x \cos x - 1)$$

$$y'' \geq 0 \quad \Rightarrow \quad (\sin x + \cos x)(3\sin x \cos x - 1) \geq 0 \quad \Rightarrow$$

$$\Rightarrow \frac{\pi}{2} - \alpha \leq x \leq \alpha \quad \vee \quad \frac{3}{4}\pi \leq x \leq \frac{3}{2}\pi \quad \vee \quad \alpha + \pi \leq x \leq \frac{7}{4}\pi,$$

$$\alpha = \arctan\left(\frac{3+\sqrt{5}}{2}\right) = 69°05'$$

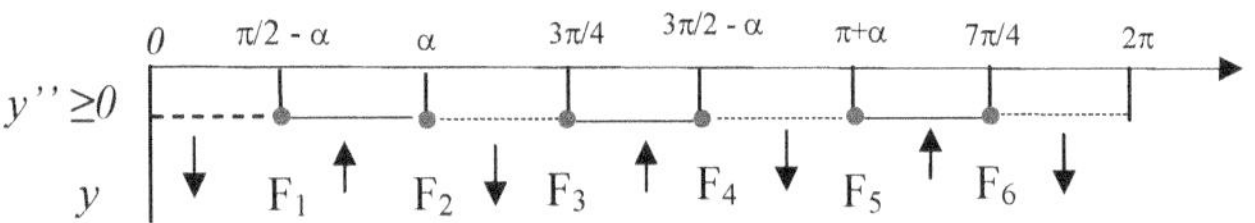

Inflection points

$$F_1\left(\frac{\pi}{2} - \alpha; \frac{2\sqrt{15}}{9}\right), \ F_2\left(\alpha; \frac{2\sqrt{15}}{9}\right), \ F_3\left(\frac{3\pi}{4};0\right), \ F_4\left(\frac{3\pi}{2} - \alpha; -\frac{2\sqrt{15}}{9}\right), \ F_5\left(\alpha + \pi; -\frac{2\sqrt{15}}{9}\right), \ F_6\left(\frac{7\pi}{4};0\right)$$

$$y\left(\frac{\pi}{2} - \alpha\right) = sen^3\left(\frac{\pi}{2} - \alpha\right) + \cos^3\left(\frac{\pi}{2} - \alpha\right) = \left[sen\left(\frac{\pi}{2} - \alpha\right)\right]^3 + \left[\cos\left(\frac{\pi}{2} - \alpha\right)\right]^3 = \left[\cos\alpha\right]^3 + \left[sen\alpha\right]^3 = \dots$$

The graph of function is represented in Figure 4.2.

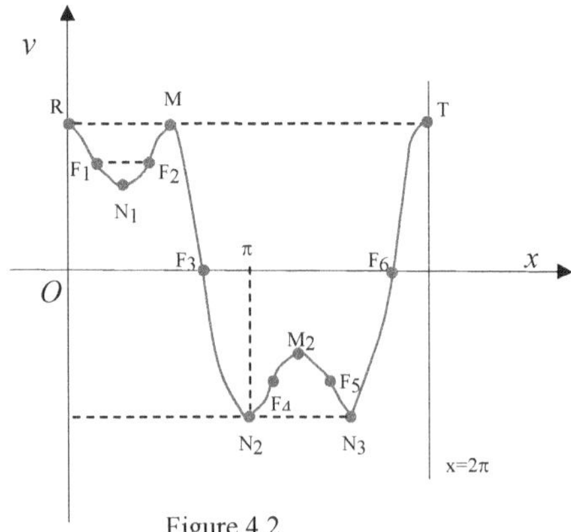

Figure 4.2

N.3.- Graph the function $y = x + 2\sin x$. $\forall x \in [-2\pi,\ 2\pi]$.

DOMAIN *D*

$$D = R$$

SYMMETRIE AND PERIODICITIES

$$y(-x) = -x + 2\sin(-x) = -x - 2\sin x = -(x + 2\sin x) = -y(x)\ x \Rightarrow$$

$$\Rightarrow \text{ y is a odd function. }^{[3]}$$

Graph the function $y \quad \forall x \in \left[0,\ 2\pi\right]$

INTERSECTION OF THE GRAPH WITH THE COORDINATE AXES

$$\begin{cases} y = x + 2senx \\ x = 0 \end{cases} \Rightarrow \begin{cases} y = 0 \\ x = 0 \end{cases} \Rightarrow O(0,0)$$

$$x = 2\pi \Rightarrow y = 2\pi \Rightarrow A(2\pi,\ 2\pi).$$

SIGN OF FUCTION

$$y \geq 0 \quad \forall x \in [0,2\pi]\ ^{[4]}$$

THE FIRST ORDER DERIVATIVE

$$y' = 1 + 2\cos x.$$

$$y' \geq 0 \Rightarrow 1 + 2\cos x \geq 0 \Rightarrow \cos x \geq -\frac{1}{2} \Rightarrow 0 \leq x \leq \frac{2}{3}\pi \ \vee \ \frac{4}{3}\pi \leq x \leq 2\pi$$

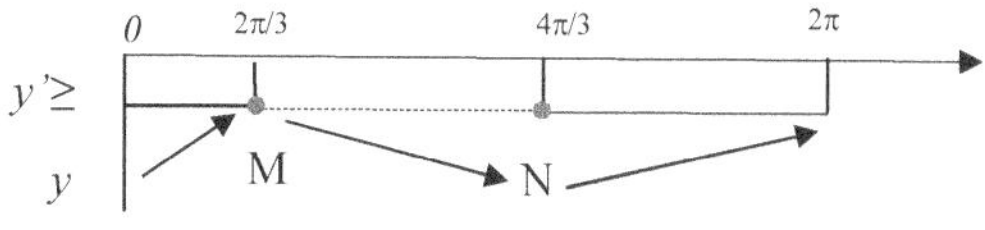

---

[3] $sin(-x) = -\sin x$.

[4] $x \geq 2senx \quad \forall x \in [0,2\pi]$ www.matematicus.com

Local minimums and maximums: $\quad M\left(\dfrac{2}{3}\pi\,;\dfrac{2}{3}\pi+\sqrt{3}\right),\quad N\left(\dfrac{4}{3}\pi\,;\dfrac{4}{3}\pi-\sqrt{3}\right).$

$$y\left(\frac{2}{3}\pi\right)=\frac{2}{3}\pi+2sen\left(\frac{2}{3}\pi\right)=\frac{2}{3}\pi+2sen\left(\frac{\pi}{3}\right)=\frac{2}{3}\pi+2\frac{\sqrt{3}}{2}=\frac{2}{3}\pi+\sqrt{3}$$

THE SECOND ORDER DERIVATIVE

$y''=-2\sin x.$

$y''\geq 0 \quad\Rightarrow\quad \sin x\leq 0 \quad\Rightarrow\quad \pi\leq x\leq 2\pi \qquad \Rightarrow\ y \text{ is convex } \forall x\in\,]\pi,\,2\pi[$

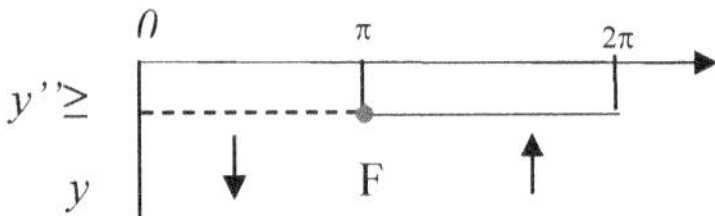

Inflection point: $F(\pi,\pi)$

The graph of function is represented in Figure 4.3

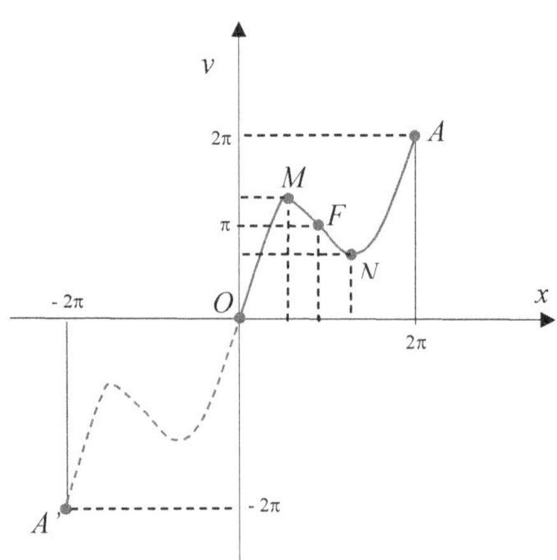

Figure 4.3

N.4.- Graph the function $y = \sin\left(x + \dfrac{\pi}{3}\right) + \cos\left(\dfrac{\pi}{6} - x\right)$ $\quad \forall x \in \left[-\dfrac{\pi}{3}, \dfrac{5}{3}\pi\right]$.

DOMAIN D

$$D = R$$

$$y = \sin\left(x + \frac{\pi}{3}\right) + \cos\left(\frac{\pi}{6} - x\right) = \sin\left(x + \frac{\pi}{3}\right) + \sin\left(x + \frac{\pi}{3}\right) = 2\sin\left(x + \frac{\pi}{3}\right)$$

$$\cos\left(x - \frac{\pi}{6}\right) = \cos\left(\frac{\pi}{6} - x\right) = \sin\left[\frac{\pi}{2} - \left(\frac{\pi}{6} - x\right)\right] = \sin\left(x + \frac{\pi}{3}\right)$$

SYMMETRIE AND PERIODICITIES

$$y(x + 2\pi) = 2\sin\left(x + 2\pi + \frac{\pi}{3}\right) = 2\sin\left((x + 2\pi) + \frac{\pi}{3}\right) =$$

$$= 2\left[\sin(x + 2\pi)\cos\frac{\pi}{3} + \cos(x + 2\pi)sen\frac{\pi}{3}\right] = \qquad \Rightarrow$$

$$= 2\left[\sin(x)\cos\frac{\pi}{3} + \cos(x)\sin\frac{\pi}{3}\right] = 2\left[\sin\left(x + \frac{\pi}{3}\right)\right] = y(x)$$

$$\Rightarrow y \text{ is a periodic function, period } p = 2\pi.$$

Graph the function $y = 2\sin\left(x + \dfrac{\pi}{3}\right)$ $\quad \forall x \in [0, 2\pi]$

$$Y = sen\ X \ \rightarrow: \begin{cases} y = 2Y \\ x = X - \dfrac{\pi}{3} \end{cases}$$

The graph of function is represented in Figure 4.4

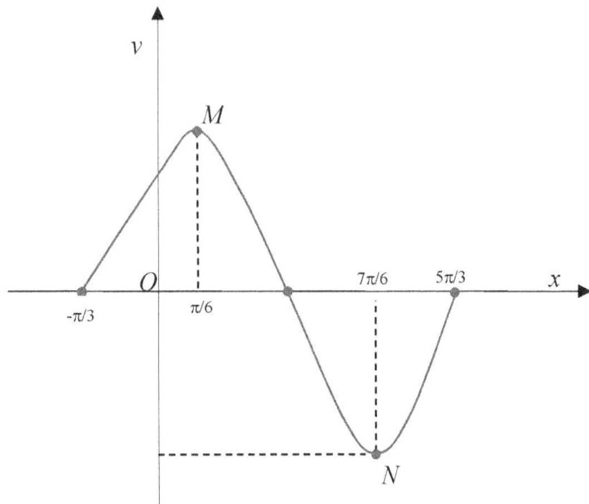

Figure 4.4

N.5.- Graph the function $y = \tan x - 2x \quad \forall x \in [-\pi, \pi]$.

DOMAIN D

$$\tan x = \frac{\sin x}{\cos x}$$

$$\downarrow$$

$$\cos x \neq 0 \;\Rightarrow\; x \neq \pi/2 + k\pi \quad \forall k \in Z \;\Rightarrow\; D = [-\pi, \pi] - \{-\pi/2, \pi/2\}.$$

$$y(\pi) = \tan(\pi) - 2(\pi) = 0 - 2\pi = -2\pi \;\Rightarrow\; A(\pi, -2\pi).$$

$$y(-\pi) = 2\pi \;\Rightarrow\; A'(\pi, 2\pi).$$

SYMMETRIE AND PERIODICITIES

$$y(-x) = \operatorname{tg}(-x) - 2(-x) = -\operatorname{tg} x + 2x = -(\operatorname{tg} x - 2x) = -y(x) \;\Rightarrow$$

$$\Rightarrow\; y \text{ is a odd function}[5]$$

Graph the function $y = \tan x - 2x \quad \forall x \in [0, \pi] - \left\{\dfrac{\pi}{2}\right\}$

INTERSECTION OF THE GRAPH WITH THE COORDINATE AXES

$$\begin{cases} y = \tan x - 2x \\ x = 0 \end{cases} \;\Rightarrow\; \begin{cases} y = 0 \\ x = 0 \end{cases} \;\Rightarrow\; O(0,0)$$

LIMITS AND ASYMPTOTES

$$\begin{cases} \lim\limits_{x \to \frac{\pi}{2}^-} (\tan x - 2x) = +\infty \\ \lim\limits_{x \to \frac{\pi}{2}^+} (\tan x - 2x) = -\infty \end{cases} \;\Rightarrow\; x = \frac{\pi}{2} \text{ vertical asymptote}$$

---

[5] $\tan(-x) = -\tan x$,

THE FIRST ORDER DERIVATIVE

$$y' = \frac{1}{\cos^2 x} - 2 = \frac{1 - 2\cos^2 x}{\cos^2 x}$$

$$y' \geq 0 \quad \Rightarrow \quad 1 - 2\cos^2 x \geq 0 \quad \Rightarrow \quad \frac{\pi}{4} \leq x \leq \frac{3}{4}\pi$$

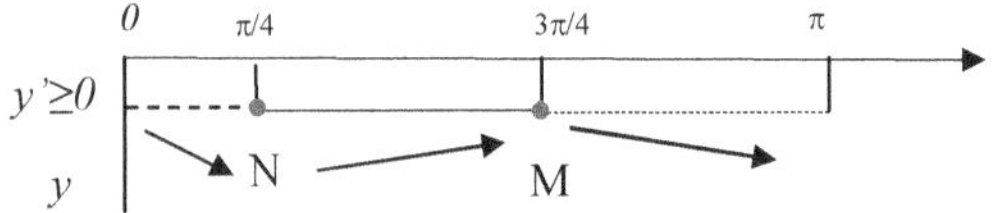

Local minimums and maximums:

$$M\left(\frac{3}{4}\pi, -1 - \frac{3}{2}\pi\right), \quad N\left(\frac{\pi}{4}, 1 - \frac{\pi}{2}\right)$$

THE SECOND ORDER DERIVATIVE

$$y'' = \frac{2\operatorname{sen} x}{\cos^3 x}, \quad \Rightarrow \quad F(0,0) \text{ Inflection point.}$$

The graph of function is represented in Figure 4.5.

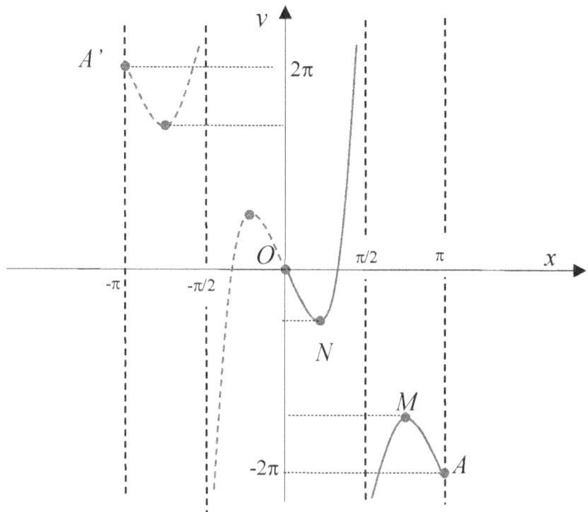

Figure 4.5

www.matematicus.com

Giulio D. Broccoli, *The graph of a function* – www.matematicus.com

N.6.- Graph the function $y = \sin x + \dfrac{1}{2}\sin 2x + \dfrac{1}{2}\sin 3x$.

DOMAIN D

$$D = R$$

SYMMETRIES AND PERIODICITIES

$$y(2\pi) = \sin(2\pi) + \frac{1}{2}\sin 2(2\pi) + \frac{1}{2}\sin 3(2\pi) = y(x) \;\Rightarrow$$

$$\Rightarrow \; y \text{ is a periodic function, period } p = 2\pi,$$

$$y(-x) = \sin(-x) + \frac{1}{2}\sin 2(-x) + \frac{1}{2}\sin 3(-x) = -y(x) \;\Rightarrow$$

$$\Rightarrow \; y \text{ is a odd function}$$

Graph the function y $\quad \forall x \in [0,\,\pi]$

SIGN OF FUCTION

$y \geq 0 \;\Rightarrow$

$$\sin x + \frac{1}{2}\sin 2x + \frac{1}{3}\sin 3x \geq 0 \;\Rightarrow\; \sin x + \sin x \cos x + \frac{1}{3}\left(3\sin x - 4\sin^3 x\right) \geq 0 \;\Rightarrow^6$$

$$\Rightarrow\; \sin x\left(\frac{4}{3}\cos^2 x + \cos x + \frac{2}{3}\right) \geq 0 \;\Rightarrow$$

$$\Rightarrow\; \begin{cases} \sin x \geq 0 \\ \dfrac{4}{3}\cos^2 x + \cos x + \dfrac{2}{3} \geq 0 \end{cases} \cup \begin{cases} \sin x \leq 0 \\ \dfrac{4}{3}\cos^2 x + \cos x + \dfrac{2}{3} \leq 0 \end{cases} \Rightarrow \forall x \in [0,\pi]$$

$$x = 0 \;\Rightarrow\; y = 0$$

$$x = \pi \;\Rightarrow\; y = 0$$

THE FIRST ORDER DERIVATIVE

$$y' = \cos x + \frac{1}{2}2\cos 2x + \frac{1}{3}3\cos 3x = \cos x + \cos 2x + \cos 3x$$

$$y' \geq 0 \;\Rightarrow\; \cos x + \cos 2x + \cos 3x \geq 0 \;\Rightarrow$$

---

[6] $\sin 3x = 3\sin x - 4\sin^3 x$

$$\Rightarrow \quad \cos x + \cos^2 x - \operatorname{sen}^2 x + 4\cos^3 x - 3\cos x \geq 0 \quad \Rightarrow^7$$

$$\Rightarrow \quad \cos x + \cos^2 x - 1 + \cos^2 x + 4\cos^3 x - 3\cos x \geq 0 \quad \Rightarrow$$

$$\Rightarrow \quad 4\cos^3 x + 2\cos^2 x - 2\cos x - 1 \geq 0 \quad \Rightarrow$$

$$\downarrow$$

$$4t^3 + 2t^2 - 2t - 1 \geq 0 \Rightarrow \left(t + \frac{1}{2}\right)\!\left(4t^2 - 2\right) \geq 0, \quad \cos x = t$$

$$\downarrow$$

$$\Rightarrow \quad 2(2\cos^2 x - 1)\left(\cos x + \frac{1}{2}\right) \geq 0 \quad \Rightarrow \quad (2\cos^2 x - 1)(2\cos x + 1) \geq 0 \quad \Rightarrow$$

$$\Rightarrow \quad 0 \leq x \leq \frac{\pi}{4} \ \lor \ \frac{2\pi}{3} \leq x \leq \frac{3\pi}{4}$$

$$x = \pi/4 \ \Rightarrow \ y' = 0$$
$$x = 2\pi/3 \ \Rightarrow \ y' = 0$$
$$x = 3\pi/4 \ \Rightarrow \ y' = 0$$

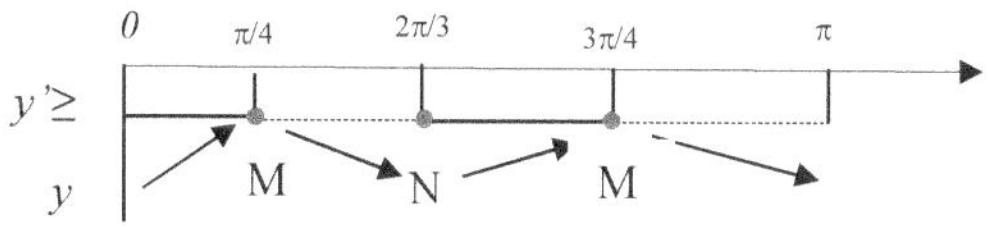

Local minimums and maximums:

$$M_1\left(\frac{\pi}{4}; \frac{4\sqrt{2}+3}{6}\right), \ N\left(\frac{2}{3}\pi; \frac{\sqrt{3}}{4}\right), \ M_2\left(\frac{3}{4}\pi; \frac{4\sqrt{2}-3}{6}\right).$$

$$y\left(\frac{\pi}{4}\right) = \operatorname{sen}\left(\frac{\pi}{4}\right) + \frac{1}{2}\operatorname{sen}2\left(\frac{\pi}{4}\right) + \frac{1}{3}\operatorname{sen}3\left(\frac{\pi}{4}\right) = \frac{\sqrt{2}}{2} + \frac{1}{2} + \frac{1}{3}\left[3\operatorname{sen}\left(\frac{\pi}{4}\right) - 4\operatorname{sen}^3\left(\frac{\pi}{4}\right)\right] =$$

$$= \frac{\sqrt{2}}{2} + \frac{1}{2} + \operatorname{sen}\left(\frac{\pi}{4}\right) - \frac{4}{3}\operatorname{sen}^3\left(\frac{\pi}{4}\right) = \frac{\sqrt{2}}{2} + \frac{1}{2} + \frac{\sqrt{2}}{2} - \frac{4}{3}\left(\frac{\sqrt{2}}{2}\right)^3 =$$

$$= \frac{\sqrt{2}}{2} + \frac{1}{2} + \frac{\sqrt{2}}{2} - \frac{4}{3}\frac{2\sqrt{2}}{8} = \frac{\sqrt{2}}{2} + \frac{1}{2} + \frac{\sqrt{2}}{3} = \frac{4\sqrt{2}+3}{6}.$$

$^7 \cos 3x = 4\cos^3 x - 3\cos x$

THE SECOND ORDER DERIVATIVE

$$y'' = -\sin x - 2\sin 2x - 3\sin 3x = -2\sin x(6\cos^2 x + 2\cos x - 1)$$

$$y'' \geq 0 \quad \Rightarrow \quad -2\sin x(6\cos^2 x + 2\cos x - 1) \geq 0 \quad \Rightarrow$$
$$\Rightarrow \quad 6\cos^2 x + 2\cos x - 1 \leq 0 \ \lor \ x = 0, x = \pi \quad \Rightarrow \quad x = 0 \lor x = \pi \lor \alpha \leq x \leq \beta$$

$$\alpha = \arccos\left(\frac{-1-\sqrt{7}}{6}\right), \beta = \arccos\left(\frac{-1+\sqrt{7}}{6}\right).$$

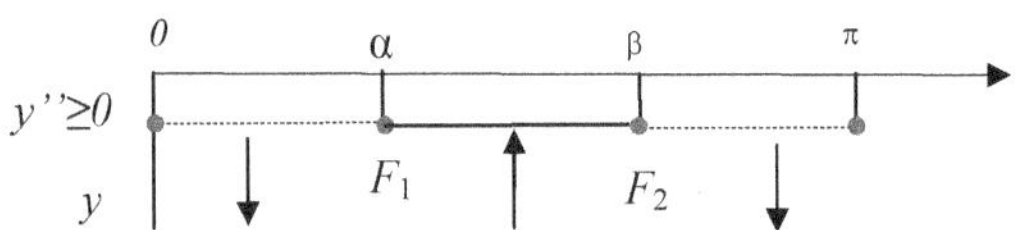

Inflection points: O(0,0), F_1, F_2

The graph of function is represented in Figure 4.6,

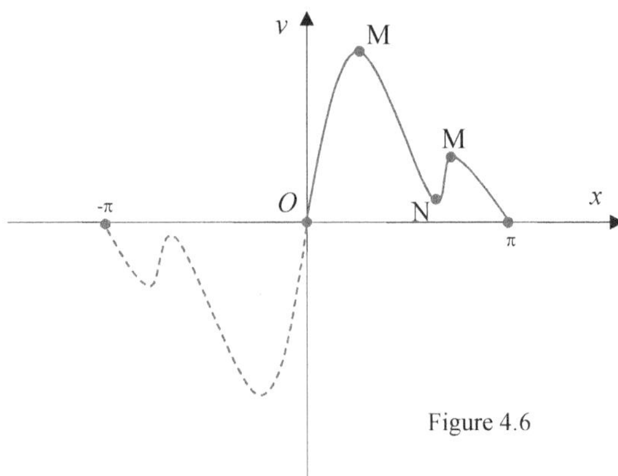

Figure 4.6

N.7.- Graph the function $y = \arcsin\left(x^2 + \dfrac{1}{2}\right)$.

DOMAIN D

$$-1 \le x^2 + \frac{1}{2} \le 1 \quad \Rightarrow \quad \begin{cases} x^2 + \dfrac{1}{2} \le 1 \\[2mm] x^2 + \dfrac{1}{2} \ge -1 \end{cases} \quad \Rightarrow \quad -\frac{\sqrt{2}}{2} \le x \le \frac{\sqrt{2}}{2} \quad \Rightarrow D = \left[-\frac{\sqrt{2}}{2}, \frac{\sqrt{2}}{2} \right]$$

SYMMETRIES AND PERIODICITIES

$$y(-x) = \arcsin\left[(-x)^2 + \frac{1}{2} \right] = \arcsin\left(x^2 + \frac{1}{2} \right) = y(x) \quad \Rightarrow$$

$$\Rightarrow \quad \text{is a even function}$$

$$\textbf{Graph the function } y = \arcsin\left(x^2 + \frac{1}{2} \right) \quad \forall x \in D_1 = \left[0, \frac{\sqrt{2}}{2} \right]$$

SIGN OF FUCTION

$$y \ge 0 \quad \forall x \in D_1$$

THE FIRST ORDER DERIVATIVE

$$y' = \frac{2x}{\sqrt{1 - \left(x^2 + \dfrac{1}{2} \right)^2}}, \quad \forall x \in \left] -\frac{\sqrt{2}}{2}, \frac{\sqrt{2}}{2} \right[$$

$$y' \ge 0 \quad \Rightarrow \quad \frac{2x}{\sqrt{1 - \left(x^2 + \dfrac{1}{2} \right)^2}} \ge 0 \quad 2x \ge 0 \quad \Rightarrow \quad x \ge 0 \quad \Rightarrow \quad 0 \le x < \frac{\sqrt{2}}{2}$$

$$\lim_{x \to \frac{\sqrt{2}}{2}^-} y' = +\infty$$

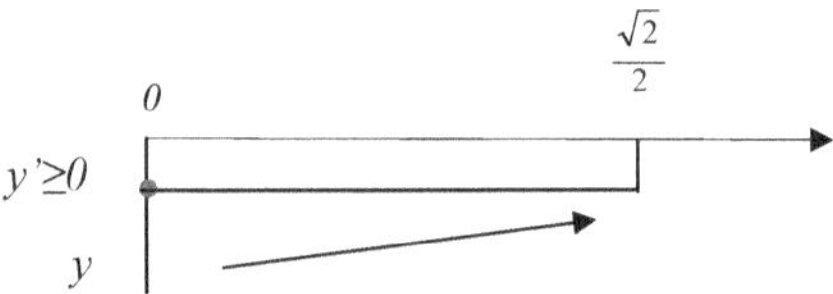

Local minimum: $N(0,\pi/6)$

The graph of function is represented in Figure 4.7.

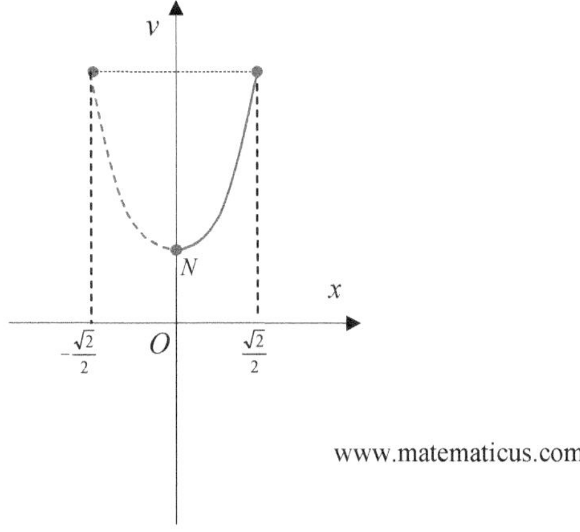

Figure 4.7

N.8.- Graph the function $y = x + \arcsin x$

DOMAIN D

$$D = [-1,1]$$

SYMMETRIES AND PERIODICITIES

$$y(-x) = -x + \arcsin(-x) = -x - \arcsin x = -(x + \arcsin x) = -y(x) \Rightarrow$$

$$\Rightarrow y \text{ is a odd function } [8]$$

Graph the function $y = x + \arcsin x \quad \forall x \in D_1 = [0, 1]$

SIGN OF FUCTION

$y \geq 0 \quad \forall x \in [0,1]$ [9]

THE FIRST ORDER DERIVATIVE

$$y' = 1 + \frac{1}{\sqrt{1-x^2}}, \quad \forall x \in [0,1[$$

$$y' = 1 + \frac{1}{\sqrt{1-x^2}} > 0 \quad \Rightarrow \quad 0 \leq x < 1$$

$$\Rightarrow \quad y \text{ is increasing } \forall x \in D_1$$

$$\lim_{x \to 1^-} y' = +\infty$$

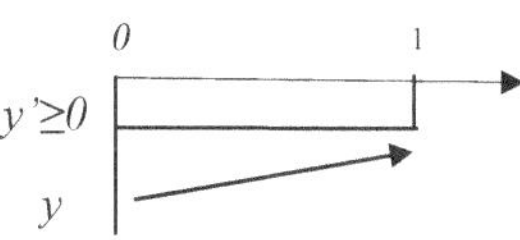

---

[8] $\arcsin x$ is a odd function.

[9] $\forall x > 0 \ \arcsin x > 0 \Rightarrow x + \arcsin x > 0.$

THE SECOND ORDER DERIVATIVE

$$y'' = \frac{x}{\left(1 - x^2\right)\sqrt{1 - x^2}} \quad , \ \forall x \in \left[0,1\right[$$

$$y'' \geq 0 \Rightarrow \frac{x}{\left(1 - x^2\right)\sqrt{1 - x^2}} \geq 0 \quad \Rightarrow \quad 0 \leq x < 1 \qquad \Rightarrow y \text{ is convex } \forall x \in \left]0, 1\right[$$

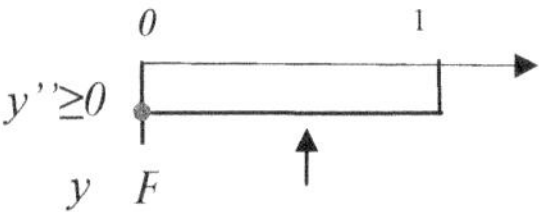

Inflection point: $F(0;0)$

The graph of function is represented in Figure 4.8.

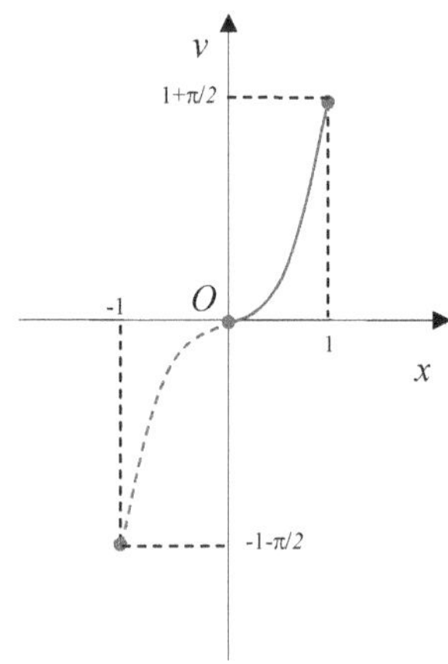

Figure 4.8

www.matematicus.com

N.9.- Graph the function $y = \arctan\left(\dfrac{x^2}{x-1}\right)$.

DOMAIN D

$$\begin{cases} \dfrac{x^2}{x-1} \in R \\[2mm] x-1 \neq 0 \end{cases} \Rightarrow \quad x-1 \neq 0 \quad x \neq 1 \quad \Rightarrow \quad D = R \text{-} \{1\}.$$

SIGN OF FUCTION

$$y \geq 0 \quad \Rightarrow \quad \arctan\left(\dfrac{x^2}{x-1}\right) \geq 0 \quad \Rightarrow \quad \dfrac{x^2}{x-1} \geq 0 \quad \Rightarrow \quad x = 0 \vee x > 1 \quad {}^{10}$$

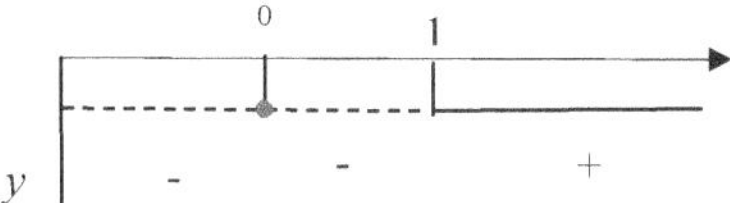

$O(0,0$

LIMITS AND ASYMPTOTES

$$\lim_{x \to 1^+} \arctan\left(\dfrac{x^2}{x-1}\right) = \dfrac{\pi}{2}, \quad \lim_{x \to 1^-} \arctan\left(\dfrac{x^2}{x-1}\right) = -\dfrac{\pi}{2} \Rightarrow x = 1 \text{ jump discontinuity.}$$

$$\lim_{x \to \pm\infty} \arctan\left(\dfrac{x^2}{x-1}\right) = \arctan\left[\lim_{x \to \pm\infty}\left(\dfrac{x^2}{x-1}\right)\right] = \arctan(\pm\infty) = \pm\infty,$$

$$m = \lim_{x \to \pm\infty} f'(x) = \lim_{x \to \pm\infty} \dfrac{x^2 - 2x}{(x-1)^2 + x^4} = \lim_{x \to \pm\infty} \dfrac{x^2}{x^4} = \lim_{x \to \pm\infty} \dfrac{1}{x^2} = 0$$

THE FIRST ORDER DERIVATIVE

$$y' = \dfrac{1}{1 + \left(\dfrac{x^2}{x-1}\right)^2} \cdot \dfrac{2x(x-1) - x^2}{(x-1)^2} = \dfrac{x^2 - 2x}{(x-1)^2 + x^4}$$

---

[10] $\arctan x > 0 \Leftrightarrow x > 0$

$$y' \geq 0 \Rightarrow \frac{x^2 - 2x}{(x-1)^2 + x^4} \geq 0 \quad \Rightarrow \quad x^2 - 2x \geq 0 \quad \Rightarrow \quad x \leq 0 \vee x \geq 2$$

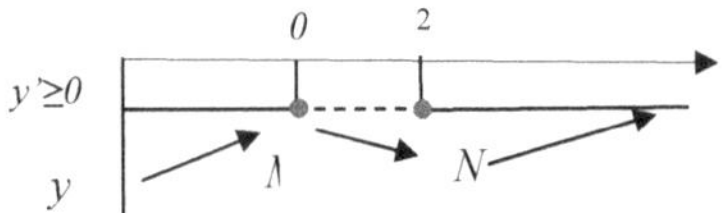

Local minimums and maximums: $M(0,0)$, $N(\,2\,;\, \arctan 4\,)$

The graph of function is represented in Figure 4.9.

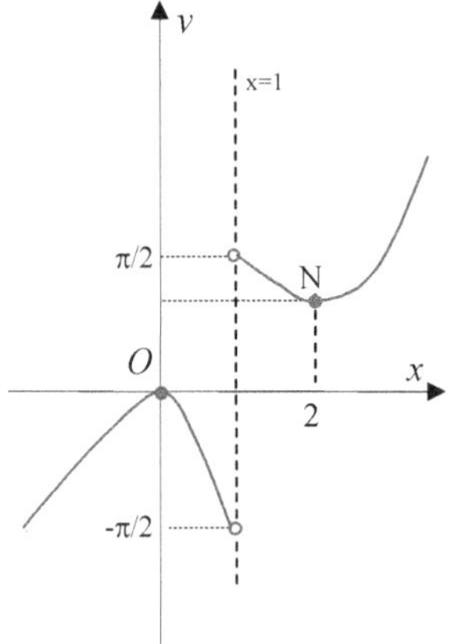

Figure 4.9

N.10.- Graph the function $y = x^2 + arctan\, x$

DOMAIN D

$$D = R.$$

SIGN OF FUNCTION

$$y \geq 0 \quad \Rightarrow \quad x^2 + \arctan x \geq 0 \quad \Rightarrow \quad x^2 \geq -\arctan x \Rightarrow x < \alpha \approx -0{,}85, \lor x > 0$$

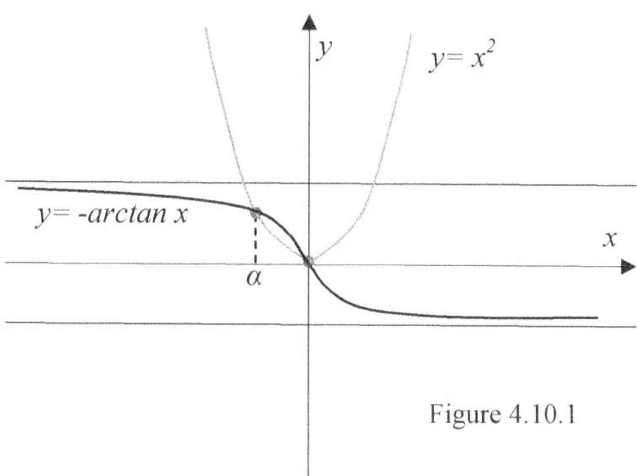

Figure 4.10.1

x	x^2	$- arctan\, x$	$x^2 + (arctan\, x)$
-1	1	0,78	1 - 0,78 = 0,22
-0,8	0,64	0,67	0,64 - 0,67 = -0,03
α = **-0,85**	0,72	0,704	0,72 - 0,704 = 0,016
-0,84	0,705	0,69	0,015
Table. 1			

$$x = \alpha \quad \Rightarrow \quad \alpha^2 + \arctan \alpha = 0$$

LIMITS AND ASYMPTOTES

$$\lim_{x \to \pm\infty} y = +\infty,$$

$$m = \lim_{x \to \pm\infty} f'(x) = \lim_{x \to \pm\infty} 2x + \frac{1}{1+x^2} = 2(+\infty) + \frac{1}{+\infty} = +\infty + 0 = +\infty$$

THE FIRST ORDER DERIVATIVE

$$y' = 2x + \frac{1}{1+x^2}$$

$$y' \geq 0 \quad \Rightarrow \quad 2x + \frac{1}{1+x^2} \geq 0 \quad \Rightarrow \quad \frac{2x^3 + 2x + 1}{1+x^2} \geq 0 \Rightarrow 2x^3 + 2x + 1 \geq 0$$

$$\Rightarrow x + b \geq 0 \qquad \Rightarrow x \geq b \approx -0,424$$

Local minimum: $x = b \approx -0,424$.

The graph of function is represented in Figure 4.10

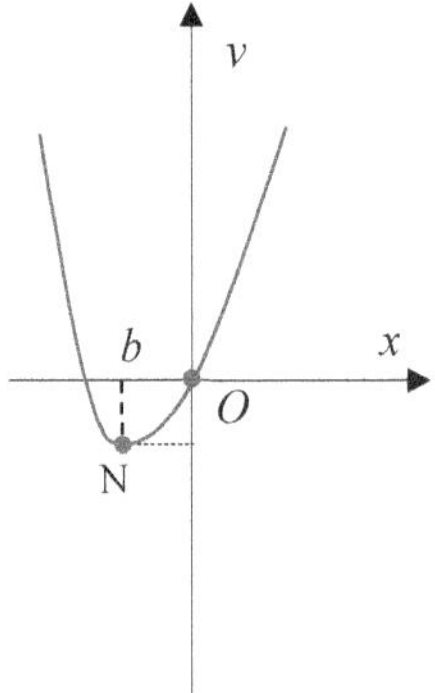

Figure 4.10

Exercises.- Graph the functions.

a) $y = \sin x + \sqrt{3}\cos x - 2$, $x \in [-\pi, \pi]$;

b) $y = 3\cos 2x - 4\cos x$, $x \in [0, 2\pi]$

c) $y = \sin(2x)\cos x$, $x \in [0, 2\pi]$;

d) $y = \sin x + \dfrac{1}{4\sin x}$, $x \in]-\pi, \pi[$

e) $y = \arctan(x^2 - 1)$

f) $y = \sin^2 x$,

g) $y = 2\sin x + \sin 2x$,

h) $y = \arctan x - x/2$,

i) $y = 3\cos x - x/2$,

l) $y = (x-1)\cos x$, $x \in [0, 2\pi]$,

m) $y = x + \sin 2x$, $x \in [-\pi, \pi]$,

n) $y = \sin 2x + 2\cos x$,

o) $y = 2\sin x + \sin(2x + \dfrac{\pi}{2})$,

p) $y = \dfrac{\sin x - \cos x}{\sin x + \cos x}$,

q) $y = \arctan\sqrt{x-1}$,

r) $y = x^2 + \arccos x$,

s) $y = x + \arctan\left(\dfrac{x^3 + x}{2}\right)$,

t) $y = \dfrac{1 - \sin x}{1 - \cos x}$.

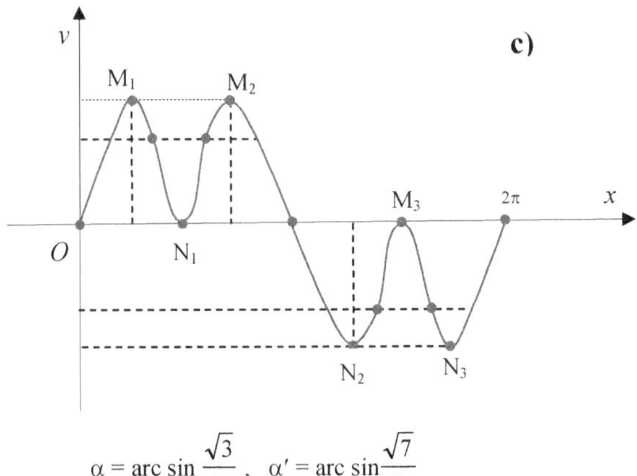

$$\alpha = \arcsin \frac{\sqrt{3}}{3} \ , \quad \alpha' = \arcsin \frac{\sqrt{7}}{3}$$

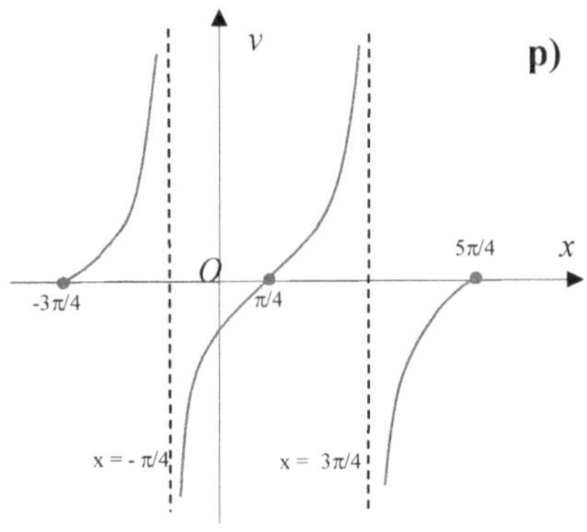

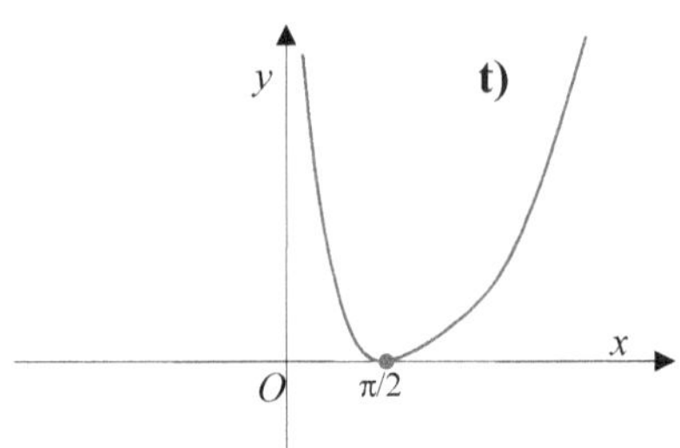

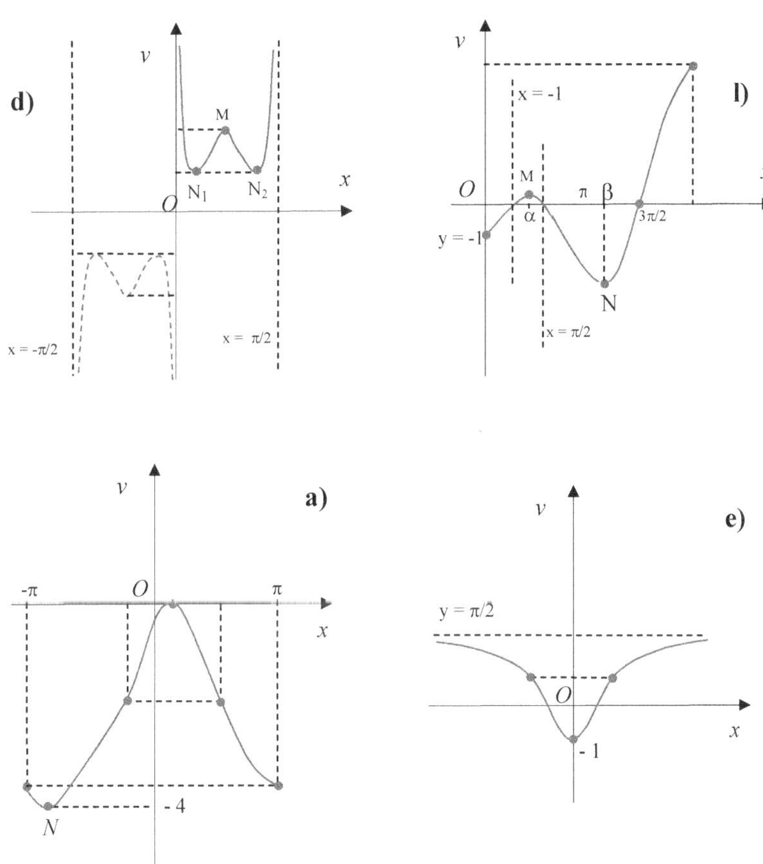

d)
v
M
N₁
N₂
x
O
x = -π/2
x = π/2
l)
v
x = -1
O
M
α
π
β
3π/2
x
y = -1
N
x = π/2
a)
v
-π
O
π
x
- 4
N
e)
v
y = π/2
O
x
- 1

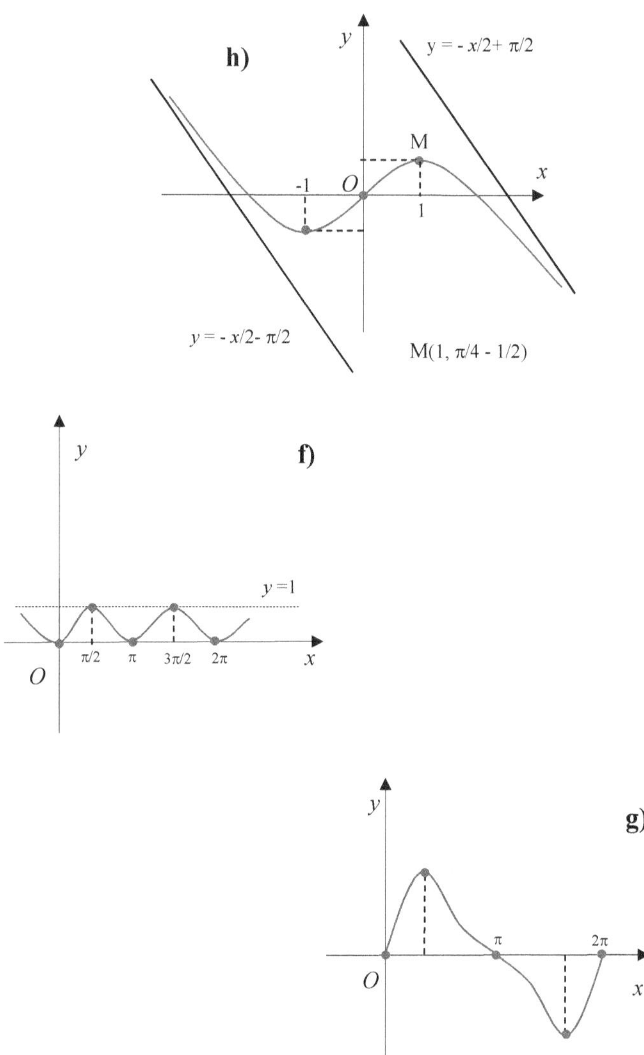
h)
y
y = - x/2 + π/2
M
-1
O
x
1
y = - x/2 - π/2
M(1, π/4 - 1/2)
f)
y
y = 1
π/2   π   3π/2   2π
x
O
g)
y
π
2π
O
x

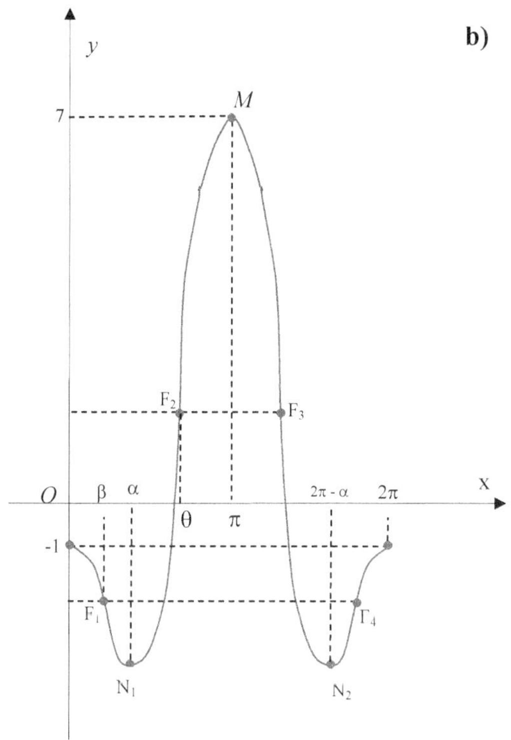

$$\beta = \arccos \frac{1+\sqrt{73}}{12} \ , \ \alpha = \arccos (1/3) \ , \ \theta = \arccos \frac{1-\sqrt{73}}{12}.$$

www.matematicus.com

5. Graphs of exponential and logarithmic functions.

N.1.- Graph the function $y = xe^{-x}$.

DOMAIN

$$D = R$$

SIGN OF FUCTION

$$y \geq 0 \quad \Rightarrow \quad xe^{-x} \geq 0 \quad \Rightarrow \quad \frac{x}{e^x} \geq 0 \quad \Rightarrow \quad x \geq 0^{[1]}$$

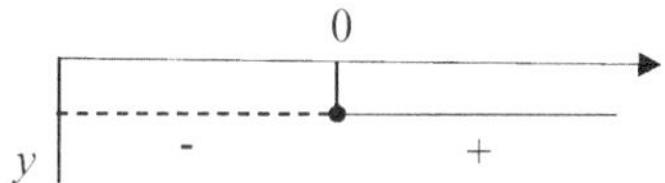

$$x = 0 \quad \Rightarrow \quad y = 0 \quad \Rightarrow \quad O(0,0)$$

LIMITS AND ASYMPTOTES

$$\lim_{x \to +\infty} xe^{-x} = \lim_{x \to +\infty} \frac{x}{e^x} \overset{DH}{=} \lim_{x \to +\infty} \frac{1}{e^x} = 0 \quad \Rightarrow y = 0 \ \text{horizontal asymptote}$$

$$\lim_{x \to -\infty} \frac{x}{e^x} = \frac{-\infty}{0^+} = -\infty$$

$$m = \lim_{x \to -\infty} \frac{y}{x} = \lim_{x \to -\infty} \frac{x}{xe^x} = \lim_{x \to -\infty} \frac{1}{e^x} = \frac{1}{0^+} = +\infty$$

THE FIRST ORDER DERIVATIVE

$$y' = \frac{1-x}{e^x}$$

$$y' \geq 0 \quad \Rightarrow \quad \frac{1-x}{e^x} \geq 0 \quad \Rightarrow \quad 1-x \geq 0 \quad \Rightarrow \quad x \leq 1$$

---

[1] $e^x > 0 \quad \forall x \in R.$

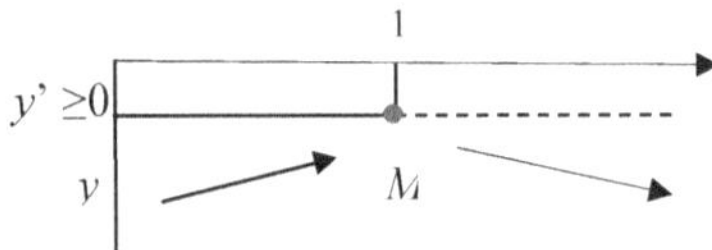

Local maximum: $M(1; 1/e)$, $\quad y(1) = 1e^{-1} = 1/e$

THE SECOND ORDER DERIVATIVE

$$y'' = \frac{x-2}{e^x}$$

$$y'' \geq 0 \quad \Rightarrow \quad \frac{x-2}{e^x} \geq 0 \quad \Rightarrow \quad x-2 \geq 0 \quad \Rightarrow \quad x \geq 2$$

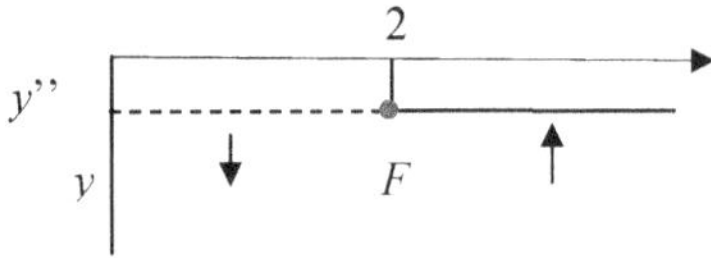

Inflection point: $F\left(2, 2e^{-2}\right)$, $\quad y(2) = 2e^{-2}$

The graph of function is represented in Figure 5.1

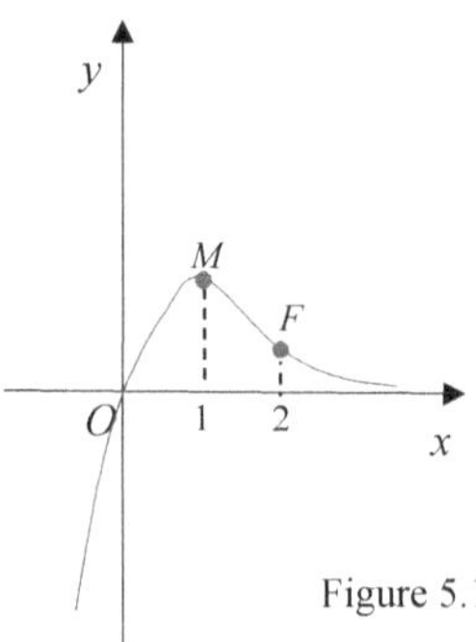

Figure 5.1

N.2.- Graph the function $y = ln(x^2 - 1)$.

DOMAIN D

$$x^2 - 1 > 0 \quad \Rightarrow \quad x > 1, \ x < -1 \quad \Rightarrow \quad D =]-\infty \, ;- 1[\, \cup \,]1; +\infty [.$$

SYMMETRIES AND PERIODICITIES

$$y(-x) = ln[(-x)^2 - 1] = ln(x^2 - 1) = y(x) \ \Rightarrow y \text{ is a even function}$$

Graph the function $y = \ln(x^2 - 1) \quad \forall x \in D_1 = \,]1, +\infty[$

SIGN OF FUCTION

$$y \geq 0 \ \Rightarrow \ ln(x^2 -1) \geq 0 \ \Rightarrow \ ln(x^2 -1) \geq ln\ 1 \ \Rightarrow \ x^2 - 1 \geq 1 \ \Rightarrow \ x \geq \sqrt{2}$$

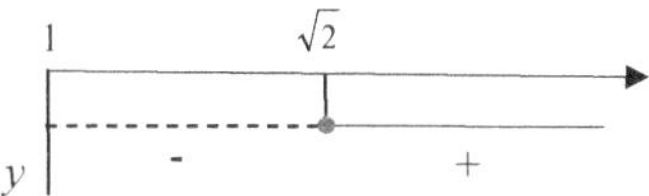

$B(\sqrt{2} \, ; 0)$.

LIMITS AND ASYMPTOTES

$$\lim_{x \to 1^+} \ln(x^2 - 1) = \log 0^+ = -\infty \ \Rightarrow \ x = 1 \ \text{vertical asymptote}$$

$$\lim_{x \to +\infty} \ln(x^2 - 1) = \ln \left| (+\infty)^2 - 1 \right| = \ln(+\infty - 1) = \ln(+\infty) = +\infty$$

$$m = \lim_{x \to +\infty} f'(x) = \lim_{x \to +\infty} \frac{2x}{x^2 - 1} = 0^{\,2}$$

$^2 \ m = \lim_{x \to \pm\infty} f'(x) \in R - \{0\}.$

THE FIRST ORDER DERIVATIVE

$$y' = \frac{2x}{x^2 - 1}$$

$$y' \geq 0 \quad \Rightarrow \quad \frac{2x}{x^2 - 1} \geq \quad \Rightarrow \quad x > 1$$

$\Rightarrow \; y$ is increasing $\forall x \in D_1$

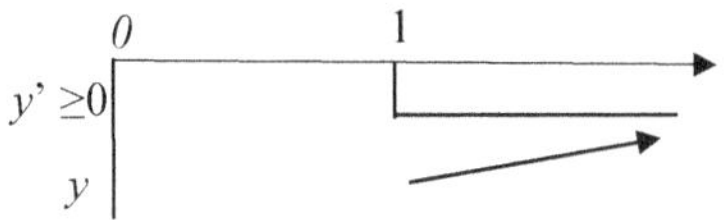

The graph of function is represented in Figure 5.2, .

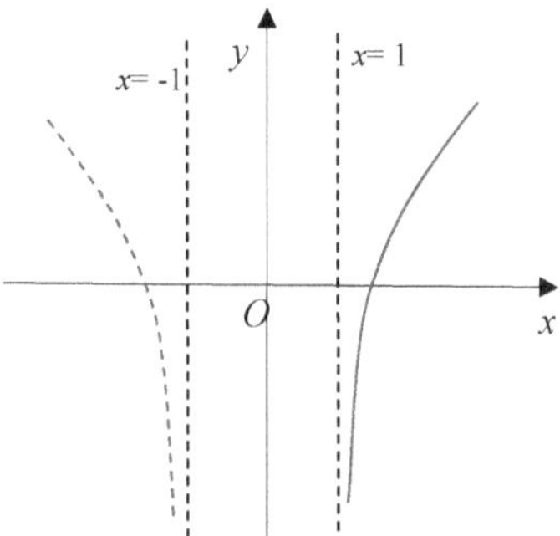

Figure 5.2

N.3.- Graph the function $y = e^{\frac{1}{1+x}}$.

DOMAIN D

$$1 + x \neq 0 \;\Rightarrow\; x \neq -1 \;\Rightarrow\; D = R - \{-1\}.$$

SIGN OF FUCTION

$y > 0 \quad \forall\, x \in D$

LIMITS AND ASYMPTOTES

$$\lim_{x \to -1^-} e^{\frac{1}{1+x}} = +\infty \;\Rightarrow\; x = -1 \;\; \text{vertical asymptote}$$

$$\lim_{x \to -1^-} e^{\frac{1}{1+x}} = e^{\frac{1}{1+(-1^-)}} = e^{\frac{1}{0^-}} = e^{-\infty} = 0 \;\Rightarrow\; P\,(-1,\,0) \;\; \text{removable discontinuity}$$

$$\lim_{x \to \pm\infty} e^{\frac{1}{1+x}} = e^{\frac{1}{1+(\pm\infty)}} = e^{\frac{1}{\pm\infty}} = e^0 = 1 = \;\Rightarrow\; y = 1 \;\; \text{horizontal asymptote}$$

THE FIRST ORDER DERIVATIVE

$$y' - e^{\frac{1}{1+x}} \frac{0 \cdot (1+x) - 1}{(1+x)^2} = - \frac{e^{\frac{1}{1+x}}}{(1+x)^2}$$

$y' \geq 0 \quad \forall x \in \varnothing \;\Rightarrow\; y \text{ is decreasing } D$

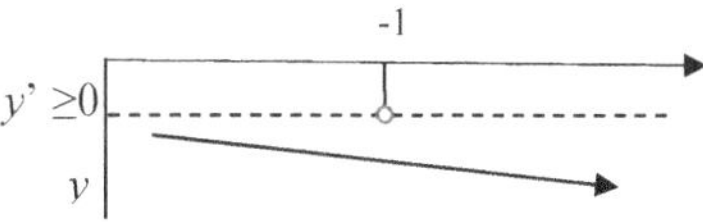

The graph of function is represented in Figure 5.3.

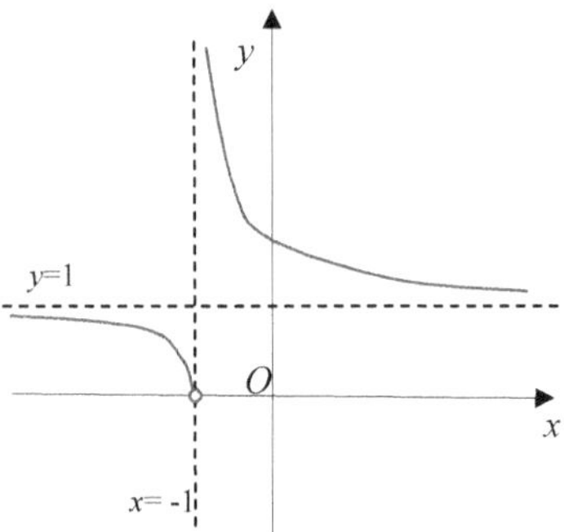

Figure 5.3

N.4.- Graph the function $y = \dfrac{\ln x}{\ln x + 1}$.

DOMAIN D

$$\begin{cases} x > 0 \\ \ln x + 1 \neq 0 \end{cases} \Rightarrow \begin{cases} x > 0 \\ \ln x \neq -1 \end{cases} \Rightarrow x > 0,\ x \neq \frac{1}{e} \Rightarrow D = \left]0, +\infty\right[- \left\{\frac{1}{e}\right\}$$

SIGN OF FUCTION

$$y \geq 0 \Rightarrow \begin{cases} \ln x \geq 0 \\ \ln + 1 > 0 \end{cases} \cup \begin{cases} \ln x \leq 0 \\ \ln x + 1 < 0 \end{cases} \Rightarrow 0 < x < \frac{1}{e} \vee x \geq 1$$

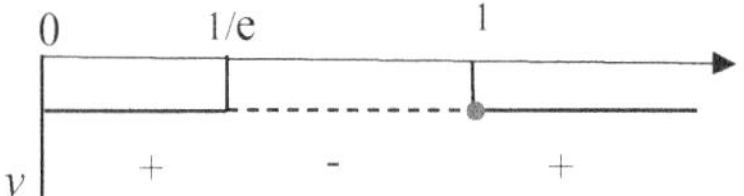

A(1;0)

LIMITS AND ASYMPTOTES

$$\lim_{x \to 0^+} \frac{\ln x}{\ln x + 1} = \lim_{x \to 0^+} \frac{\ln x}{\ln x\left(1 + \dfrac{1}{\ln x}\right)} = \lim_{x \to 0^+} \frac{1}{1 + \dfrac{1}{\ln x}} = \frac{1}{1 + \dfrac{1}{-\infty}} = \frac{1}{1 + 0} = 1^+ \Rightarrow$$

$$\Rightarrow P(0,\ 1) \text{ removable discontinuity}$$

$$\begin{cases} \lim_{x \to \frac{1}{e}^-} \dfrac{\ln x}{\ln x + 1} = \dfrac{\ln\left(\dfrac{1}{e}\right)}{\ln\left(\dfrac{1}{e}\right) + 1} = \dfrac{-1}{-1+1} = \dfrac{-1}{0^-} = +\infty \\[3em] \lim_{x \to \frac{1}{e}^+} \dfrac{\ln x}{\ln x + 1} = -\infty \end{cases} \Rightarrow x = \frac{1}{e}. \text{ vertical asymptote}$$

$$\lim_{x \to +\infty} \frac{\ln x}{\ln x + 1} = \lim_{x \to +\infty} \frac{\ln x}{\ln x\left(1 + \dfrac{1}{\ln x}\right)} = \lim_{x \to +\infty} \frac{1}{1 + \dfrac{1}{\ln x}} = \frac{1}{1 + \dfrac{1}{+\infty}} = \frac{1}{1 + 0} = 1 \Rightarrow$$

$$\Rightarrow \quad y = 1 \quad \text{horizontal asymptote}$$

THE FIRST ORDER DERIVATIVE

$$y' = \frac{\dfrac{1}{x}(\ln x + 1) - \ln x\left(\dfrac{1}{x}\right)}{(\ln x + 1)^2} = \frac{\dfrac{1}{x}\ln x + \dfrac{1}{x} - \dfrac{1}{x}\ln x}{(\ln x + 1)^2} = \frac{1}{x(\ln x + 1)^2}$$

$$y' \geq 0 \quad \Rightarrow \quad \frac{1}{x(\ln x + 1)^2} \geq 0 \quad \Rightarrow \quad \frac{1}{x} \geq 0 \quad \Rightarrow \quad x > 0$$

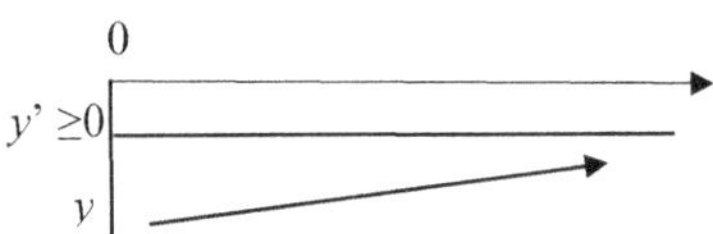

THE SECOND ORDER DERIVATIVE

$$y'' = \frac{-\left[(\ln x + 1)^2 + 2x(\ln x + 1)\dfrac{1}{x}\right]}{\left[x(\ln x + 1)^2\right]^2} = \frac{-\left(\ln^2 x + 4\ln x + 3\right)}{\left[x(\ln x + 1)^2\right]^2} = \frac{-(\ln x + 1)(\ln x + 3)}{x^2(\ln x + 1)^4}$$

$$= -\frac{3 + \ln x}{x^2(1 + \ln x)^3}$$

$$y'' \geq 0 \quad \Rightarrow \quad -\frac{3 + \ln x}{x^2(1 + \ln x)^3} \geq 0 \quad \Rightarrow \quad \frac{3 + \ln x}{1 + \ln x} \leq 0 \quad \Rightarrow \quad e^{-3} \leq x < e^{-1}$$

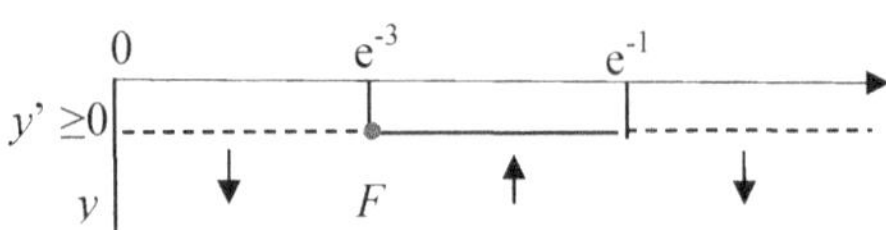

Inflection point: $F(e^{-3}\,;\,3/2\,)$.

$$y\left(e^{-3}\right) = \frac{\log e^{-3}}{\log e^{-3} + 1} = \frac{-3}{-3 + 1} = 3/2$$

The graph of function is represented in Figure .5.4.

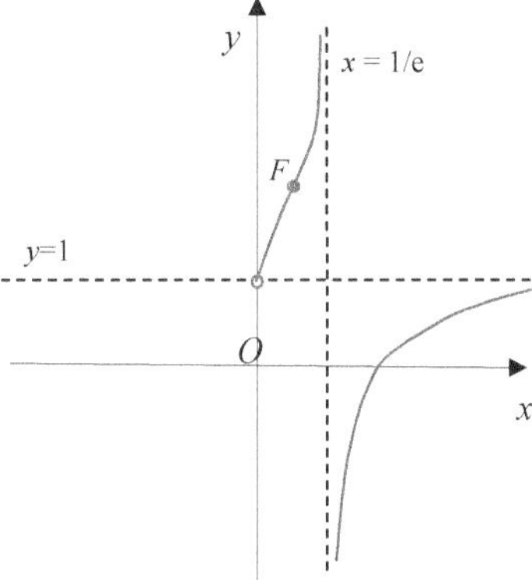

Figure 5.4

N.5.- Graph the function $y = 2^{2x} - 5 \cdot 2^x + 4$

DOMAIN D

$$D = \mathbf{R}$$

SIGN OF FUCTION

$$y \geq 0 \quad \Rightarrow \quad 2^{2x} - 5 \cdot 2^x + 4 \geq 0 \quad \Rightarrow \quad x \leq 0 \ \vee \ x \geq 2$$

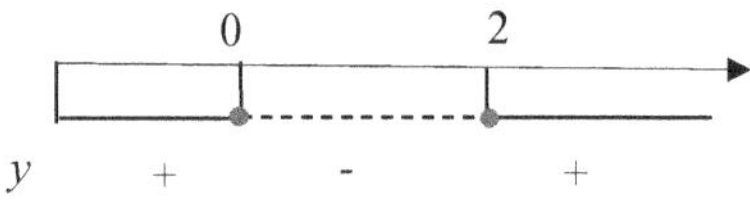

$A(2;0)$ e $O(0;0)$.

LIMITS AND ASYMPTOTES

$$\lim_{x \to -\infty} 2^{2x} - 5 \cdot 2^x + 4 = 4 \quad \Rightarrow y = 4 \ \text{ horizontal asymptote}$$

$$\lim_{x \to +\infty} 2^{2x} - 5 \cdot 2^x + 4 = +\infty$$

$$m = \lim_{x \to +\infty} \frac{2^{2x} - 5 \cdot 2^x + 4}{x} = \lim_{x \to +\infty} \frac{2^{2x}}{x} - 5 \frac{2^x}{x} + \frac{4}{x} = +\infty$$

THE FIRST ORDER DERIVATIVE

$$y' = 2^x \log 2 \cdot \left(2 \cdot 2^x - 5\right)$$

$$y' \geq 0 \quad \Rightarrow \quad 2^x \log 2 \cdot \left(2 \cdot 2^x - 5\right) \geq 0 \quad \Rightarrow \quad 2 \cdot 2^x - 5 \geq 0 \quad \Rightarrow \quad 2^x \geq \frac{5}{2} \quad \Rightarrow \quad x \geq \log_2 \frac{5}{2}$$

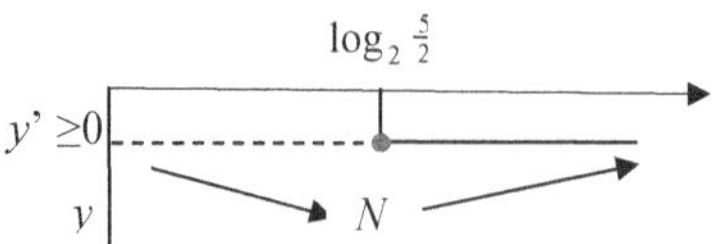

Local minimum: $N\left(\log_2 \tfrac{5}{2}; -\tfrac{9}{4}\right)$

$$y\left(\log_2 \tfrac{5}{2}\right) = 2^{2\left(\log_2 \tfrac{5}{2}\right)} - 5 \cdot 2^{\log_2 \tfrac{5}{2}} + 4 = 2^{\log_2 \left(\tfrac{5}{2}\right)^2} - 5 \cdot 2^{\log_2 \tfrac{5}{2}} + 4 = \left(\tfrac{5}{2}\right)^2 - 5 \cdot \tfrac{5}{2} + 4 = -\tfrac{9}{4}$$

THE SECOND ORDER DERIVATIVE

$$y'' = 2^x \log^2 2 \cdot \left(4 \cdot 2^x - 5\right)$$

$$y'' \geq 0 \;\Rightarrow\; 2^x \log^2 2 \cdot \left(4 \cdot 2^x - 5\right) \geq 0 \;\Rightarrow\; 4 \cdot 2^x - 5 \geq 0 \;\Rightarrow\; 2^x \geq \frac{5}{4} \;\Rightarrow\; x \geq \log_2 \frac{5}{4}$$

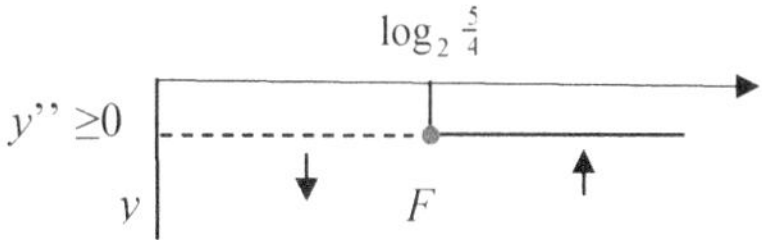

Inflection point: $F\left(\log_2 \tfrac{5}{4}; -\tfrac{11}{16}\right)$

$$y\left(\log_2 \tfrac{5}{4}\right) = 2^{2\left(\log_2 \tfrac{5}{4}\right)} - 5 \cdot 2^{\log_2 \tfrac{5}{4}} + 4 = 2^{\log_2 \left(\tfrac{5}{4}\right)^2} - 5 \cdot 2^{\log_2 \tfrac{5}{4}} + 4 = \left(\tfrac{5}{4}\right)^2 - 5 \cdot \tfrac{5}{4} + 4 = -\tfrac{11}{16}$$

The graph of function is represented in Figure . 5.5

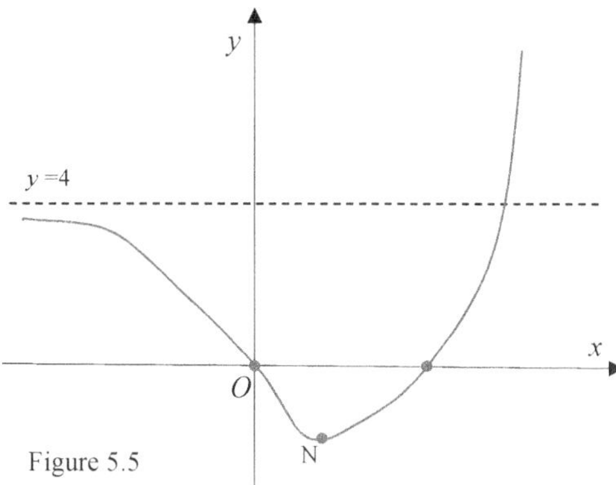

Figure 5.5

 **N.6.-** Graph the function $y = xe^{\frac{1}{x-2}}$.

DOMAIN D

$$x - 2 \neq 0 \;\Rightarrow\; x \neq 2 \;\Rightarrow\; D = R - \{2\}.$$

SIGN OF FUNCTION

$$y \geq 0 \;\Rightarrow\; xe^{\frac{1}{x-2}} \geq 0 \;\Rightarrow\; x \geq 0$$

$$x > 0 \;\Rightarrow\; y > 0$$
$$x < 0 \;\Rightarrow\; y < 0$$
$$x = 0 \;\Rightarrow\; y = 0$$

LIMITS AND ASYMPTOTES

$$\lim_{x \to 2^+} xe^{\frac{1}{x-2}} = 2e^{\frac{1}{0^+}} = 2e^{+\infty} = +\infty \;\Rightarrow\; x = 2 \text{ vertical asymptote}$$

$$\lim_{x \to 2^-} xe^{\frac{1}{x-2}} = 2e^{\frac{1}{0^-}} = 2e^{-\infty} = \frac{2}{e^{+\infty}} = 0 \;\Rightarrow\; P(2,0) \text{ removable discontinuity}$$

$$\lim_{x \to \pm\infty} xe^{\frac{1}{x-2}} = \pm\infty \;, \qquad m = \lim_{x \to \pm\infty} xe^{\frac{1}{x-2}} \cdot \frac{1}{x} = 1$$

$$n = \lim_{x \to \pm\infty} xe^{\frac{1}{x-2}} - x = (+\infty - \infty) = \lim_{x \to \pm\infty} x(e^{\frac{1}{x-2}} - 1) \overset{DH}{=} \lim_{x \to \pm\infty} \frac{e^{\frac{1}{x-2}} - 1}{\frac{1}{x}} = \lim_{x \to \pm\infty} e^{\frac{1}{x-2}} \frac{x^2}{(x-2)^2} = +1 \;\Rightarrow$$

$$\Rightarrow y = x + 1 \text{ oblique asymptote .}^{[3]}$$

THE FIRST ORDER DERIVATIVE

$$y' = e^{\frac{1}{x-2}} + xe^{\frac{1}{x-2}} \frac{-1}{(x-2)^2} = e^{\frac{1}{x-2}}\left(1 - \frac{x}{(x-2)^2}\right) = e^{\frac{1}{x-1}}\left(\frac{x^2 - 5x + 4}{(x-2)^2}\right)$$

$$y' \geq 0 \;\Rightarrow\; e^{\frac{1}{x-1}}\left(\frac{x^2 - 5x + 4}{(x-2)^2}\right) \geq 0 \;\Rightarrow\; \frac{x^2 - 5x + 4}{(x-2)^2} \geq 0 \;\Rightarrow\; x^2 - 5x + 4 \geq 0 \;\Rightarrow\; x \leq 1 \lor x \geq 4$$

---

[3] DH = L'Hôpital's rule.

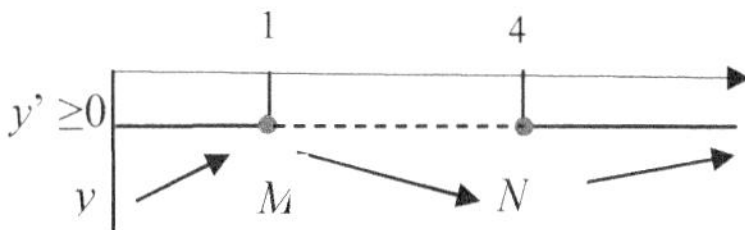

Local maxima and minima: $M(1,e^{-1})$, $N(4,4e^{-1/2})$.

The graph of function is represented in Figure .5.6.

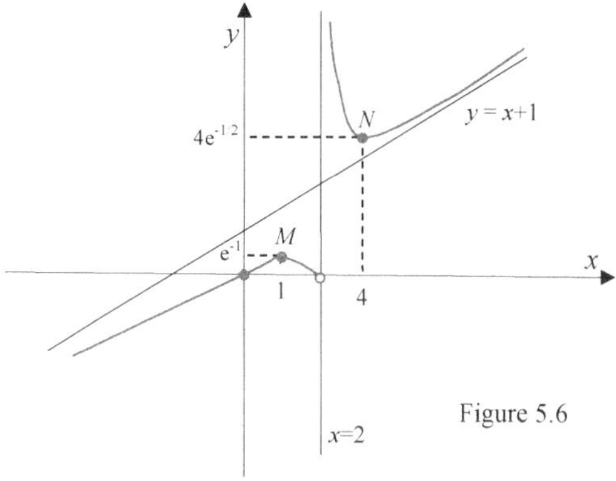

Figure 5.6

N.7.- Graph the function $y = 3x^2 + \ln(2x - 1)$.

DOMAIN *D*

$$2x - 1 > 0. \;\; \Rightarrow \; x > \tfrac{1}{2} \;\; \Rightarrow \;\; D = \,]1/2, +\infty[.$$

SIGN OF FUNCTION

$$y \geq 0 \;\; \Rightarrow \;\; 3x^2 + \ln(2x - 1) \geq 0 \;\; \Rightarrow \;\; \ln(2x - 1) \geq -3x^2$$

$$\downarrow$$

$$v(x) = \ln(2x - 1), \quad u(x) = -3x^2 \quad \text{(Figure 5.7.1)}.$$

$$\downarrow$$

$$
\begin{aligned}
x > \alpha \;\;&\Rightarrow\; y > 0, \\
x = \alpha \;\;&\Rightarrow\; y = 0 \\
\tfrac{1}{2} < x < \alpha \;\;&\Rightarrow\; y < 0
\end{aligned}
$$

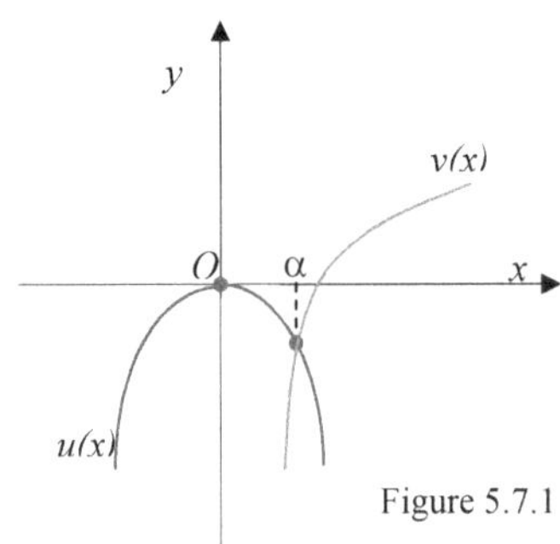

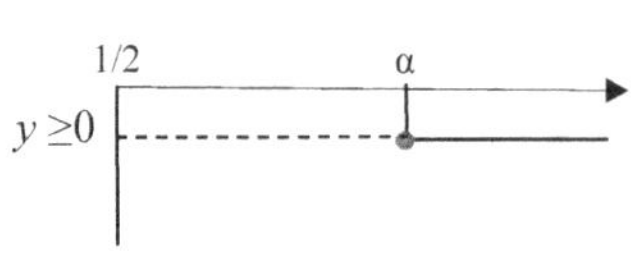

Figure 5.7.1

LIMITS AND ASYMPTOTES

$$\lim_{x \to 1/2} y = -\infty \;\; \Rightarrow \; x = \tfrac{1}{2} \;\; \text{vertical asymptote}$$

$$\lim_{x \to +\infty} y = +\infty$$

$$m = \lim_{x \to +\infty} y' = \lim_{x \to +\infty}\left(6x + \frac{2}{2x - 1} \right) = +\infty$$

THE FIRST ORDER DERIVATIVE

$$y' = 6x + \frac{2}{2x - 1} = \frac{2(6x^2 - 3x + 1)}{2x - 1}$$

$$y' \geq 0 \;\; \Rightarrow \;\; \frac{2(6x^2 - 3x + 1)}{2x - 1} \;\; \Rightarrow \;\; 6x^2 - 3x + 1 \geq 0 \;\; \Rightarrow \;\; \forall x \in R$$

THE SECOND ORDER DERIVATIVE

$$y'' = \frac{2\left(12x^2 - 12x + 1\right)}{\left(2x - 1\right)^2}$$

$$y'' \geq 0 \quad \Rightarrow \quad \frac{2\left(12x^2 - 12x + 1\right)}{\left(2x - 1\right)^2} \quad \Rightarrow \quad 12x^2 - 12x + 1 \geq 0 \Rightarrow x \leq \frac{3 - \sqrt{6}}{6} \vee x \geq \frac{3 + \sqrt{6}}{6}$$

Inflection point $x = \dfrac{3 + \sqrt{6}}{6} \approx 0{,}91 : F(0{,}91 \; ; 2{,}95)$.

The graph of function is represented in Figure 5.7.

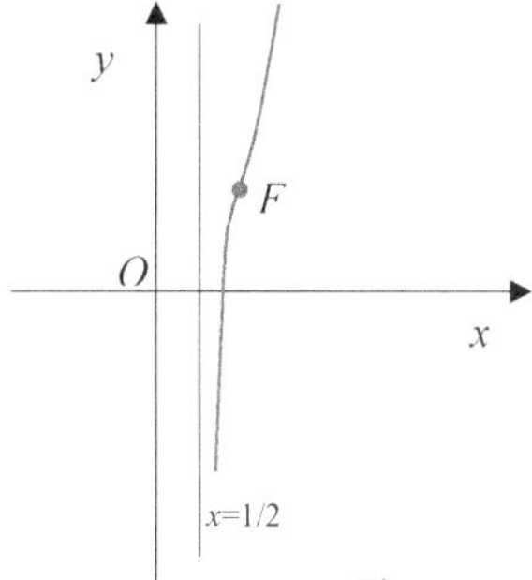

Figure 5.7

N.8.- Graph the function $y = \dfrac{\sqrt{x^2 - 4}}{\ln(x - 2)}$

DOMAIN D

$$\begin{cases} x^2 - 4 \geq 0 \\ x - 2 > 0 \\ \ln(x - 2) \neq 0 \end{cases} \Rightarrow \begin{cases} x \leq -2, \quad x \geq 2 \\ x > 2 \\ \ln(x - 2) \neq \ln 1 \end{cases} \Rightarrow \begin{cases} x \leq -2, \quad x \geq 2 \\ x > 2 \\ x - 2 \neq 1 \end{cases} \Rightarrow \begin{cases} x \leq -2, \quad x \geq 2 \\ x > 2 \\ x \neq 3 \end{cases} \Rightarrow D = \,]2, +\infty\,[- \{3\}.$$

SIGN OF FUNCTION

$$y \geq 0 \quad \Rightarrow \quad \frac{\sqrt{x^2 - 4}}{\ln(x - 2)} \geq 0 \quad \Rightarrow \quad \ln(x - 2) > 0 \quad \Rightarrow \ln(x - 2) > \ln 1 \quad \Rightarrow \quad x > 3$$

$$x > 3 \Rightarrow y > 0$$
$$2 < x < 3 \Rightarrow y < 0$$

LIMITS AND ASYMPTOTES

$$\lim_{x \to 2} \frac{\sqrt{x^2 - 4}}{\ln(x - 2)} = \frac{0}{-\infty} = 0 \quad \Rightarrow \quad x = 3 \text{ removable discontinuity}$$

$$\lim_{x \to 3^{\pm}} \frac{\sqrt{x^2 - 4}}{\ln(x - 2)} = \pm\infty \quad \Rightarrow \quad x = 3 \text{ vertical asymptote}$$

$$\lim_{x \to +\infty} \frac{\sqrt{x^2 - 4}}{\ln(x - 2)} = \left(\frac{+\infty}{+\infty}\right)^{DH} = \lim_{x \to +\infty} \frac{x^2 - 2x}{\sqrt{x^2 - 4}} = +\infty$$

$$m = \lim_{x \to +\infty} \frac{1}{x} \times \frac{\sqrt{x^2 - 4}}{\ln(x - 2)} = \lim_{x \to +\infty} \frac{\sqrt{x^2 - 4}}{x} \times \frac{1}{\ln(x - 2)} = 1 \times 0 = 0$$

THE FIRST ORDER DERIVATIVE

$$y' = \frac{2x \log(x - 2)}{2\sqrt{x^2 - 4}} - \frac{\sqrt{x^2 - 4}}{x - 2} = \frac{x(x - 2)\log(x - 2) - x^2 + 4}{(x - 2)\sqrt{x^2 - 4}}$$

$$y' \geq 0 \Rightarrow \frac{x(x - 2)\log(x - 2) - x^2 + 4}{(x - 2)\sqrt{x^2 - 4}} \geq 0 \quad \Rightarrow \quad (x - 2)\left[x \log(x - 2) - x - 2\right] \geq 0 \quad \Rightarrow$$

1) $x\ln(x-2)-x-2\geq 0 \quad \Rightarrow \quad \ln(x-2)\geq\dfrac{x+2}{x} \quad \Rightarrow \quad x>\alpha\ (\alpha>e+2),$

$$\downarrow$$

$$v(x)=\log(x-2),\quad u(x)=\dfrac{x+2}{x}\quad\text{(Figure 5.8.1)}$$

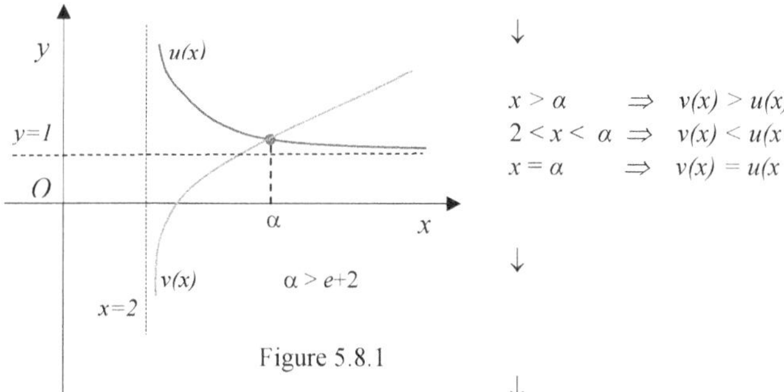

$$\downarrow$$

$x>\alpha \qquad\quad \Rightarrow\quad v(x)>u(x),$
$2<x<\alpha \Rightarrow\quad v(x)<u(x$
$x=\alpha \qquad\quad \Rightarrow\quad v(x)=u(x$

$$\downarrow$$

Figure 5.8.1

$$\downarrow$$

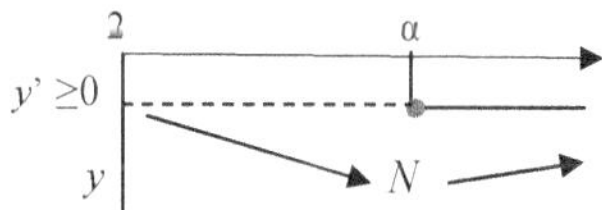

Local minima: $N(\alpha\,,\,y(\alpha))$

The graph of function is represented in Figure 5.8.

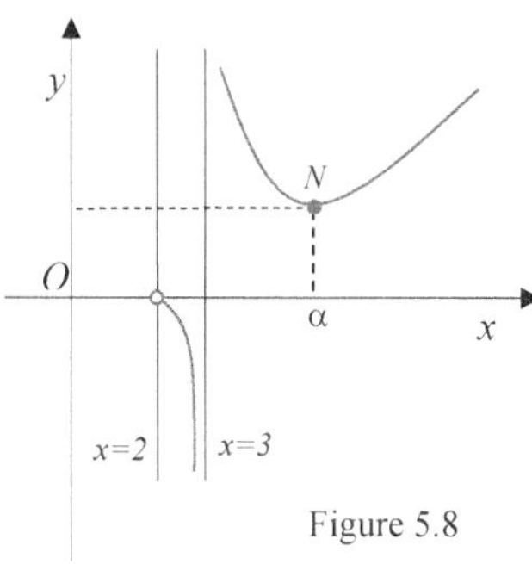

Figure 5.8

N.9.- Graph the function $y = \dfrac{e^{x^2-4}}{x-1}$.

DOMAIN D

$$x-1 \neq 0 \quad \Rightarrow \quad x \neq 1 \quad \Rightarrow \quad D = R\text{-}\{1\}.$$

SIGN OF FUNCTION

$$y \geq 0 \quad \Rightarrow \quad \frac{e^{x^2-4}}{x-1} \geq 0 \quad \Rightarrow \quad x-1 > 0 \quad \Rightarrow \quad x > 1$$

LIMITS AND ASYMPTOTES

$$\lim_{x \to 1^{\pm}} \frac{e^{x^2-4}}{x-1} = \pm\infty \quad \Rightarrow \quad x = 1 \quad \text{vertical asymptote}$$

$$\lim_{x \leftarrow \pm\infty} \frac{e^{x^2-4}}{x-1} = \left(\frac{\infty}{\infty}\right) = \lim_{x \to \pm\infty} 2x e^{x^2-4} = \pm\infty$$

$$m = \lim_{x \leftarrow \pm\infty} \frac{e^{x^2-4}}{x-1} \cdot \frac{1}{x} = \left(\frac{\infty}{\infty}\right)^{DH} = \lim_{x \to \pm\infty} e^{x^2-4} = +\infty$$

THE FIRST ORDER DERIVATIVE

$$y' = \frac{2x e^{x^2-4}(x-1) - e^{x^2-4}}{(x-1)^2} = \frac{e^{x^2-4}[2x(x-1)-1]}{(x-1)^2}$$

$$y' \geq 0 \quad \Rightarrow \quad \frac{e^{x^2-4}[2x(x-1)-1]}{(x-1)^2} \geq 0 \quad \Rightarrow \quad 2x(x-1)-1 \geq 0 \quad \Rightarrow$$

$$\Rightarrow \quad 2x^2 - 2x - 1 \geq 0 \quad \Rightarrow \quad x \leq \frac{1-\sqrt{3}}{2} \quad \vee \quad x \geq \frac{1+\sqrt{3}}{2}$$

Local minimums and maximums:

$$M: \quad x = \frac{1-\sqrt{3}}{2} \approx -0{,}36; \qquad N: x = \frac{1+\sqrt{3}}{2} \approx 1{,}36.$$

The graph of function is represented in Figure 5.9

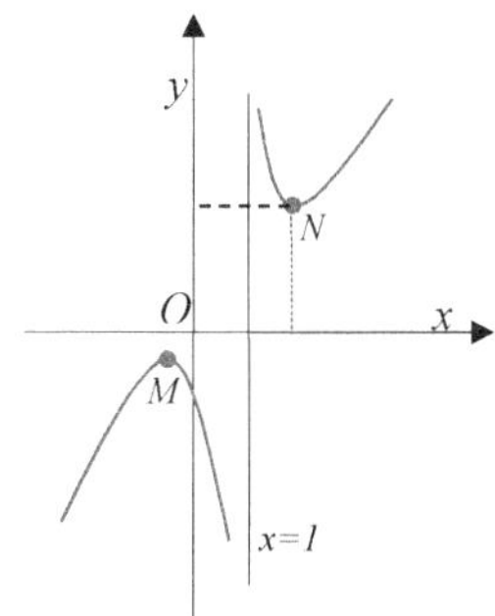

Figure 5.9

$$y(1,36) = \frac{e^{(1,36)^2-4}}{1,36-1} = \frac{e^{-2,13}}{0,36} = \frac{1}{0,36} \cdot \frac{1}{e^{2,13}} = 0,329$$

$$y(-0,36) = \frac{e^{(-0,36)^2-4}}{-0,36-1} = \frac{e^{-3,87}}{-1,36} = -\frac{1}{1,36} \cdot \frac{1}{e^{3,87}} = \dots$$

Exercises.- Graph the functions.

a) $y = \dfrac{1}{e^x(x^2 - 3)}$, b) $y = x^2 e^{1-x}$, c) $y = \dfrac{x}{\ln x}$,

d) $y = \ln\dfrac{x^2 - 2x + 1}{x + 1}$, e) $y = \begin{cases} 1 & \forall x \le 0 \\ \ln x - \ln^2 x & \forall\ x > 0 \end{cases}$, f) $y = \ln\dfrac{\sqrt{x}}{x - \sqrt{x}}$,

g) $y = e^{\frac{2x^2 - 5x}{x + 3}}$, h) $y = \ln^2(x + 3) - \ln(x + 3)$, i) $y = \dfrac{1 - e^x}{e^{2x} - 4e^x + 3}$

l) $y = e^{-\frac{x}{3}}(3x - 2)^{\frac{1}{9}}$, m) $y = e^{-\frac{x}{3}}\sqrt[9]{3x - 2}$, n) $y = \dfrac{1 - e^x}{e^{2x} - 4e^x + 3}$,

o) $y = \ln\dfrac{(x - 2)^2}{x}$, p) $y = \dfrac{x^3 - 1}{\ln(x^3 - 1)}$, q) $y = \dfrac{e^{-x}}{3 - x^2}$,

r) $y = \dfrac{e^{x-1}}{\sqrt[3]{x - 1}}$, s) $y = \dfrac{-2x^2}{\ln x}$, t) $y = e^x(1 + 4x - 3x^2)$

u) $y = 2x - 1 - \ln(e^x - 2)$, v) $y = -x\sqrt[3]{\ln x - 2}$, z) $y = 3/2 + \ln(1 + 2x^3)$,

j) $y = \ln^2 x + 5\ln x - 6$, k) $y = \ln^2 x - 2\ln x + 1$,

w) $y = \begin{cases} \dfrac{x^2}{\ln x} & \forall x > 1 \\ x^2 \ln^2 x & \forall\ 0 < x \le 1 \end{cases}$, x) $y = \begin{cases} (x + 1)^3\left[\ln(x + 1) - 1/3\right] & \forall\ x > -1 \\ x^4\left[\ln(-x) - 1/4\right] & \forall\ x \le -1 \end{cases}$

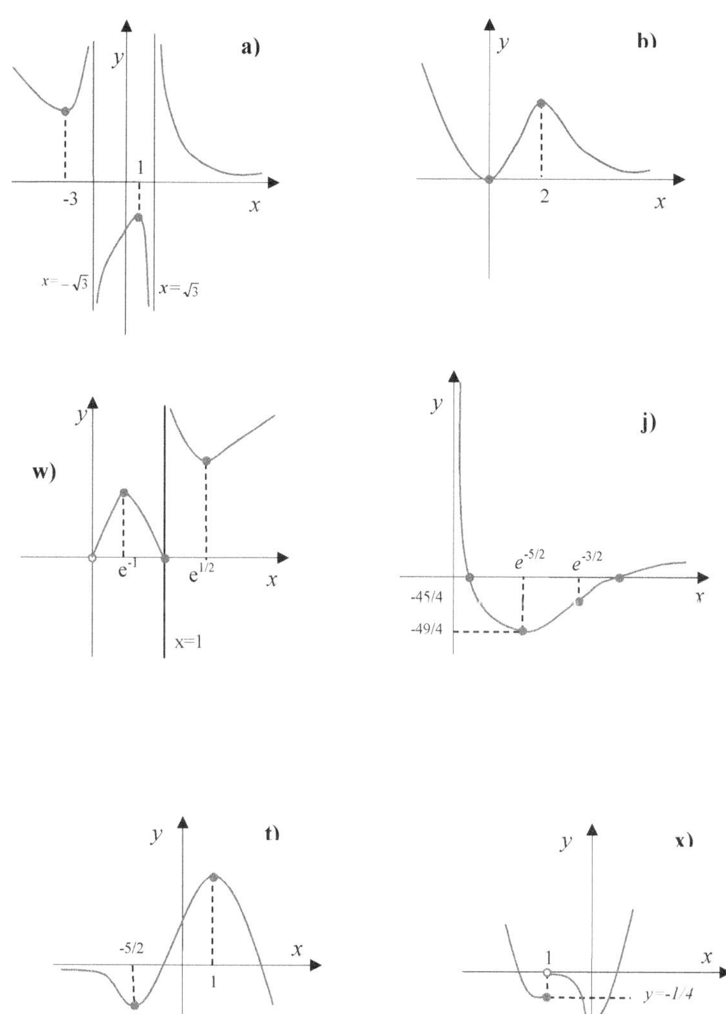
a)
y
x
-3
1
x=-√3
x=√3
b)
y
x
2
w)
y
x
e^{-1}
e^{1/2}
x=1
j)
y
e^{-5/2}
e^{-3/2}
x
-45/4
-49/4
t)
y
x
-5/2
1
x)
y
x
1
y=-1/4
y=-1/3

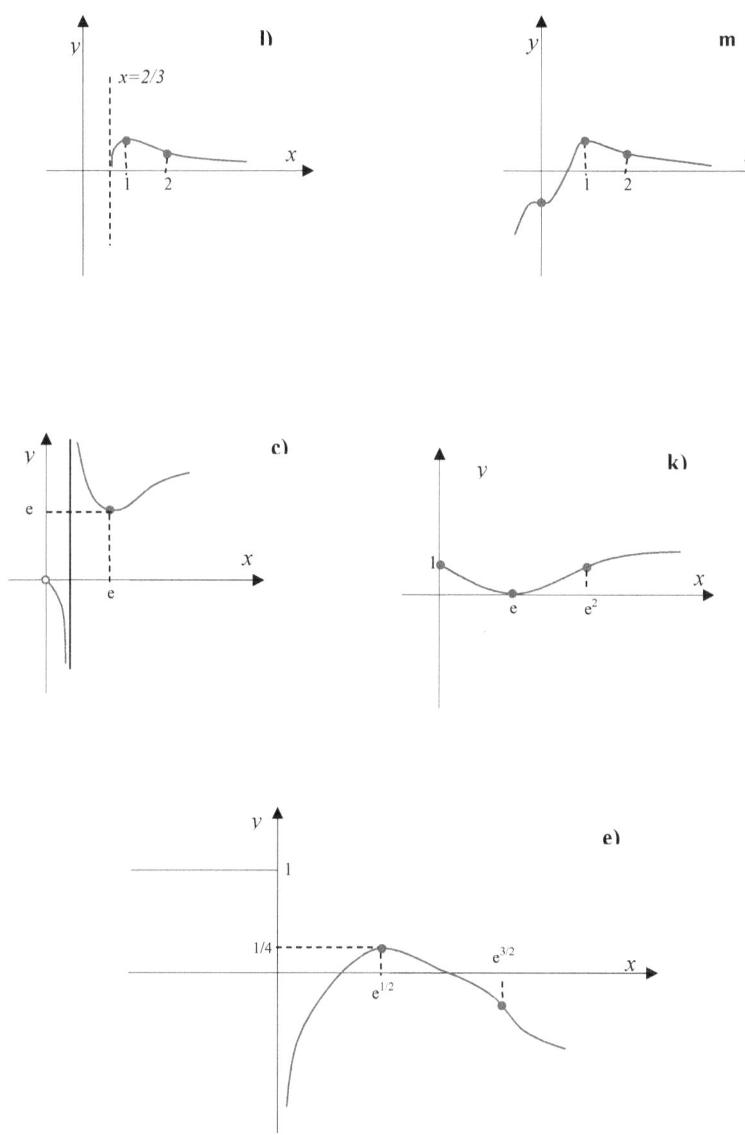

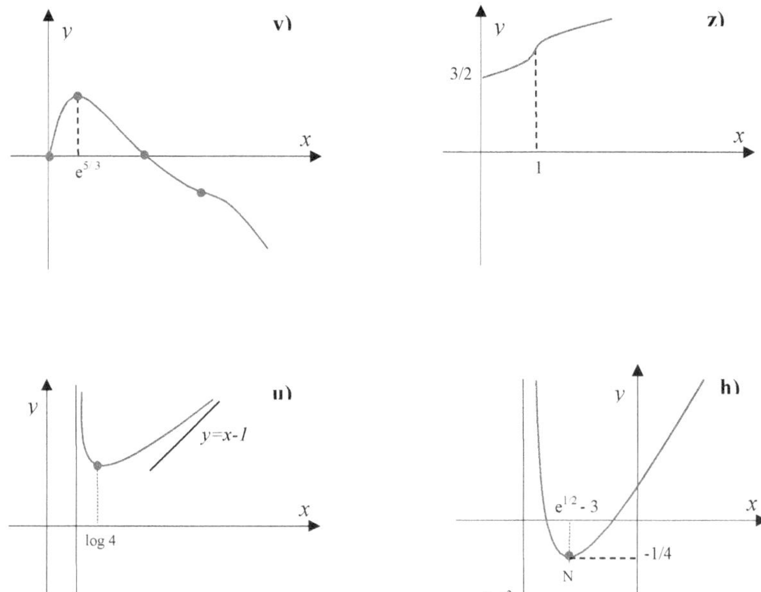
v)
y
x
e^{5/3}
z)
y
x
3/2
1
u)
y
x
y=x-1
log 4
h)
y
x
e^{1/2} - 3
-1/4
N
x=-3

6. Graphs of absolute value functions

$$y = |f(x)| \quad \Leftrightarrow \quad y = \begin{cases} f(x) & \forall x : f(x) \geq 0 \\ -f(x) & \forall x : f(x) < 0 \end{cases}$$

N.1.- Graph the function $y = |x^2 - 1|$.

$$y = |x^2 - 1| \quad \Leftrightarrow \quad y = \begin{cases} x^2 - 1 & \forall x : x^2 - 1 \geq 0 \\ -(x^2 - 1) & \forall x : x^2 - 1 < 0 \end{cases} \quad \Rightarrow$$

$$\Rightarrow \quad y = \begin{cases} x^2 - 1 & \forall x : x \geq 1, \, x \leq -1 \\ -(x^2 - 1) & \forall x : -1 < x < 1 \end{cases}$$

Graph the function $y = x^2 - 1$ (Figure 6.1)

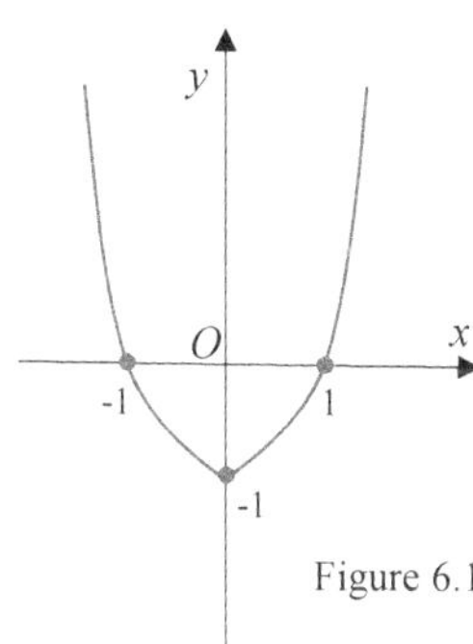

Figure 6.1

Graph the function $y = -(x^2 - 1) = 1 - x^2$ (Figure 6.2)

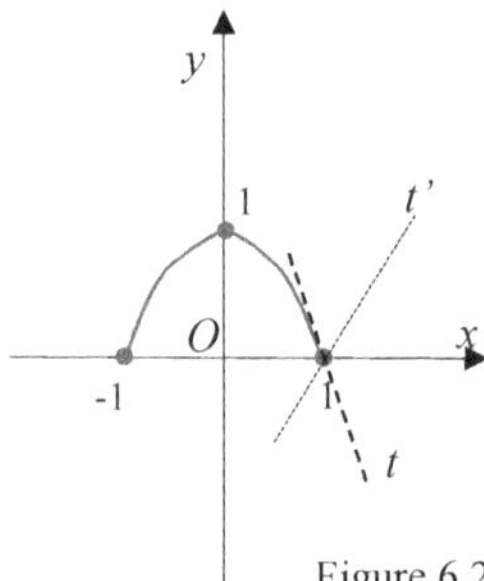

Figure 6.2

Graph the function $y = |x^2 - 1|$ (Figure 6.3)

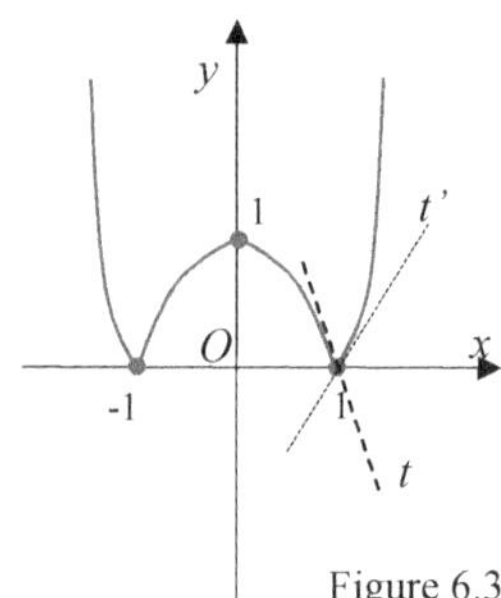

Figure 6.3

$t') \, y = 2x - 2, \qquad t) \, y = -2x - 2$

N.2.- Graph the function $y = \left| x^5 - x^2 \right|$.

$$y = \left| x^5 - x^2 \right| \quad \Leftrightarrow \quad y = \begin{cases} x^5 - x^2 & \forall x : x^5 - x^2 \geq 0 \\ -\left(x^5 - x^2\right) & \forall x : x^5 - x^2 < 0 \end{cases}$$

Graph the function $y = x^5 - x^2$ **(Figure 6.4)**

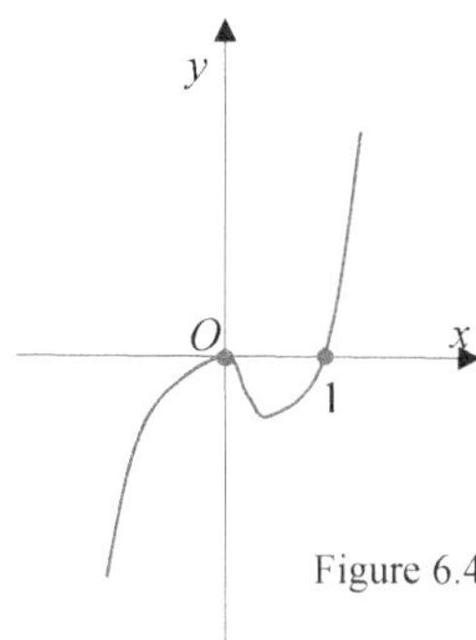

Figure 6.4

Graph the function $y = \left| x^5 - x^2 \right|$ **(Figure 6.5)**

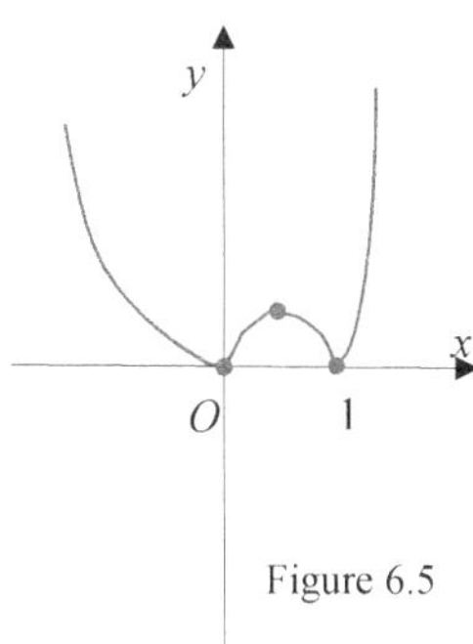

Figure 6.5

N.3.- Graph the function $y = \left| \dfrac{x^2 + 1}{x^2 - 1} \right|$.

Graph the rational function $y = \dfrac{x^2 + 1}{x^2 - 1}$ (Figure 6.6)

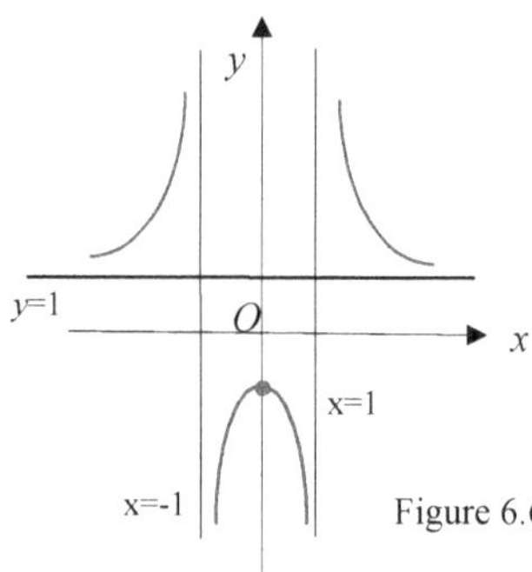

Figure 6.6

Graph the function $y = \left| \dfrac{x^2 + 1}{x^2 - 1} \right|$ (Figure 6.7)

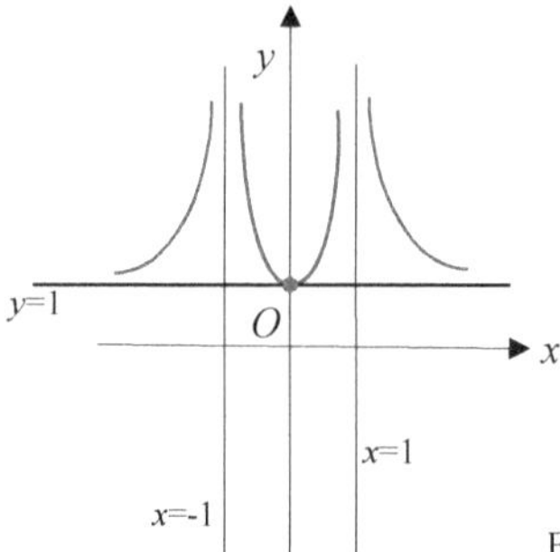

Figure 6.7

N.4.- Graph the function $y = x^3 + 2x^2 + |x|$.

$$y = x^3 + 2x^2 + |x| \quad \Leftrightarrow \quad y = \begin{cases} x^3 + 2x^2 + x & \forall\ x \geq 0 \\ x^3 + 2x^2 - x & \forall\ x < 0 \end{cases}$$

Graph the rational function $y = x^3 + 2x^2 + x \quad \forall\ x \geq 0$ (Figure 6.8)

Graph the rational function $y = x^3 + 2x^2 - x \quad \forall\ x < 0$ (Figure 6.9)

Graph the function $y = x^3 + 2x^2 + |x|$ (Figure 6.10)

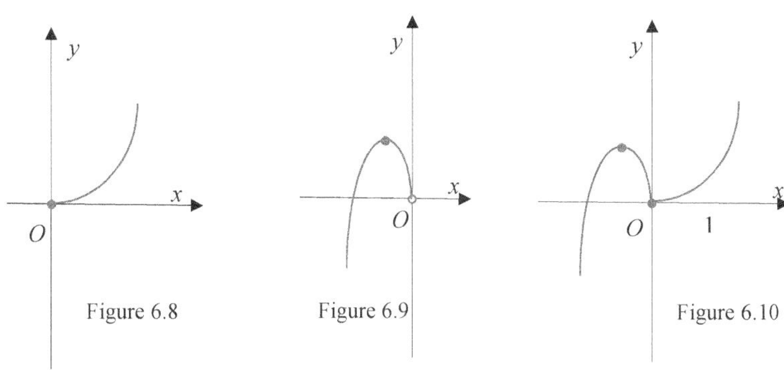

Figure 6.8   Figure 6.9   Figure 6.10

N.5.- Graph the function $y = \dfrac{x^2 + x + |x-1|}{x^2 - 1}$.

$$y = \frac{x^2 + x + |x-1|}{x^2 - 1} \iff y = \begin{cases} \dfrac{x^2 + 2x - 1}{x^2 - 1} & \forall\ x > 1 \\[3ex] \dfrac{x^2 + 1}{x^2 - 1} & \forall\ x < 1, x \neq -1 \end{cases}$$

Graph the rational function $y = \dfrac{x^2 + 2x - 1}{x^2 - 1}$ $\forall\ x > 1$ (Figure 6.11)

Graph the rational function $y = \dfrac{x^2 + 1}{x^2 - 1}$ $\forall\ x < 1, x \neq -1$ (Figure 6.12)

Graph the function $y = \dfrac{x^2 + x + |x-1|}{x^2 - 1}$ (Figure 6.13)

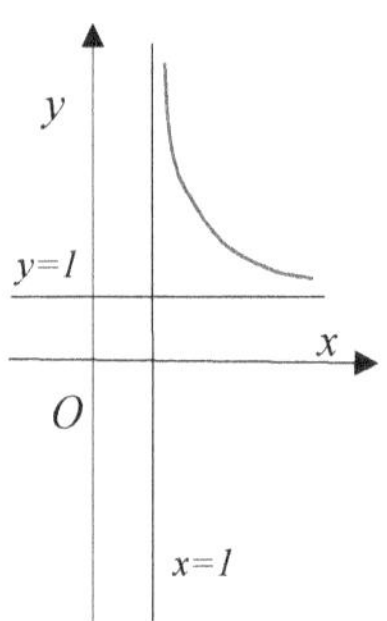

Figure 6.11

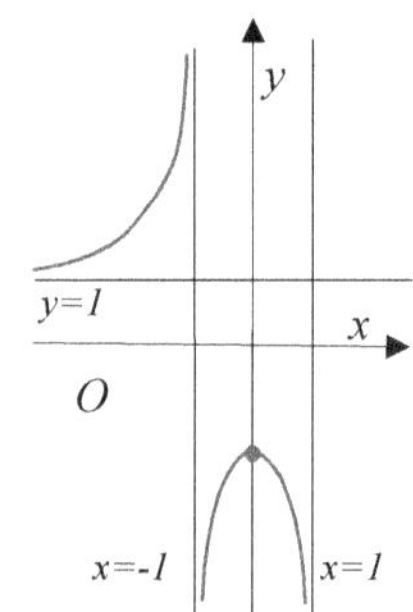

Figure 6.12

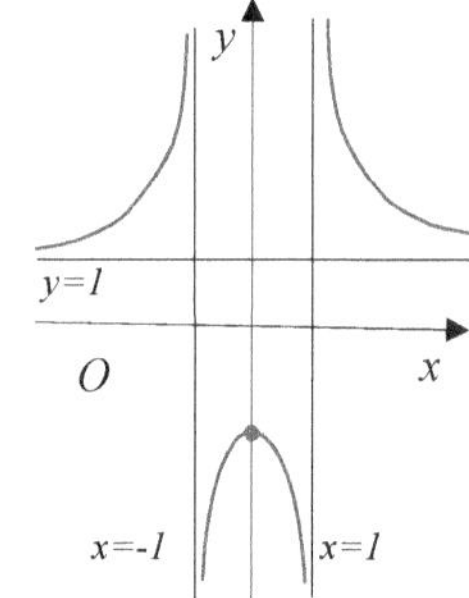

Figure 6.13

N.6.- Graph the function $y = \dfrac{4x^2}{|x-4| + |x|}$.

DOMAIN D

$$|x-4| + |x| = 0. \Rightarrow \forall x \in \varnothing \Rightarrow D = R \ ^4$$

$$y = y_1 \cup y_2 \cup y_3$$

$$y_1 = \begin{cases} x-4 \le 0 \\ x \le 0 \\ y = \dfrac{4x^2}{-x-4-x} \end{cases} \Rightarrow y_1 = \dfrac{4x^2}{-x+4-x} = \dfrac{2x^2}{2-x} \quad \forall x < 0$$

$$y_2 = \begin{cases} x-4 < 0 \\ x > 0 \\ y = \dfrac{4x^2}{-x+4+x} \end{cases} \Rightarrow y_2 = \dfrac{4x^2}{-x+4+x} = x^2 \quad \forall \, 0 \le x \le 4$$

$$y_3 = \begin{cases} x-4 > 0 \\ x > 0 \\ y = \dfrac{4x^2}{x-4+x} \end{cases} \Rightarrow y_3 = \dfrac{4x^2}{x-4+x} = \dfrac{2x^2}{x-2} \quad \forall \, x > 4$$

The graph of function $y = y_1 \cup y_2 \cup y_3$ is represented in Figure 6.14

---

^{4} $|x-4| \ge 0, \ |x| \ge 0 \Rightarrow |x-4| + |x| > 0 \ \forall x \in R$

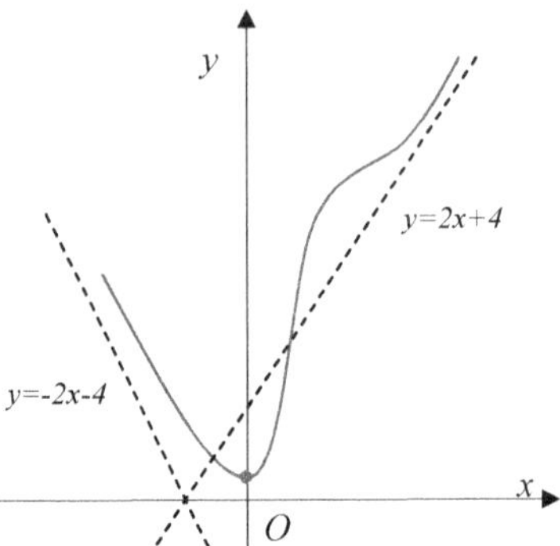

Figure 6.14

Exercises.- Graph the functions

a) $y = |x|^3 - 1$,

b) $y = |x| + 1$,

c) $y = |x^2 - 4x + 3|$,

d) $y = |x - 2| + |1 - x|$,

e) $y = |x| + |x^2 - 1|$,

f) $y = \begin{cases} \sqrt{x|x-10|} & \forall x: \ x \geq 0 \\ 1 & \forall x: \ x < 0 \end{cases}$,

h) $y = \begin{cases} \dfrac{x}{|x|+|1-x|} & \forall x: x \geq 0 \\ -x & \forall x: \ x < 0 \end{cases}$,

i) $y = |x \ln x|$,

l) $y = \left| \dfrac{e^x - 1}{e^x + 1} \right|$.

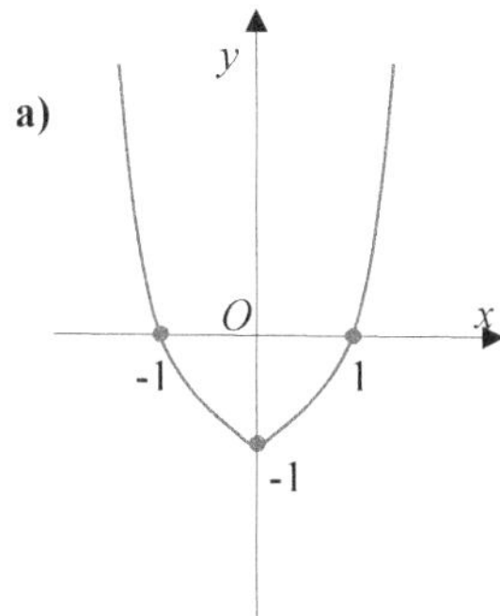

a)

$$\boxed{\textbf{a)} \quad y = f(|x|) = \begin{cases} f(x) & \forall x: \ x \geq 0 \\ f(-x) & \forall x: \ x < 0 \end{cases}}$$

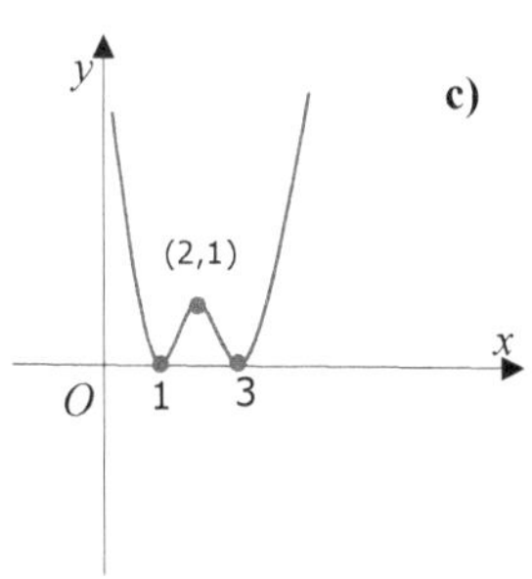

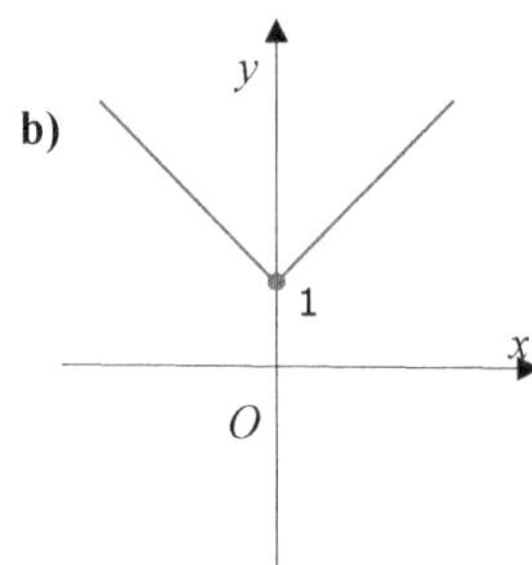

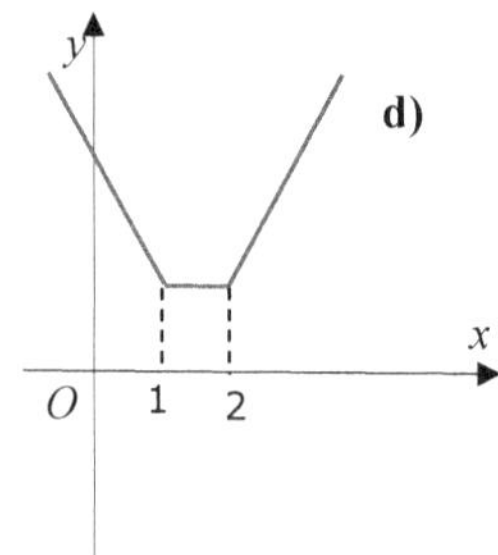

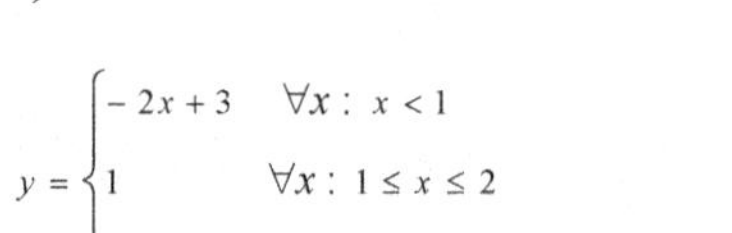

d)

$$y = \begin{cases} -2x + 3 & \forall x : x < 1 \\ 1 & \forall x : 1 \le x \le 2 \\ 2x - 3 & \forall x : x > 2 \end{cases}$$

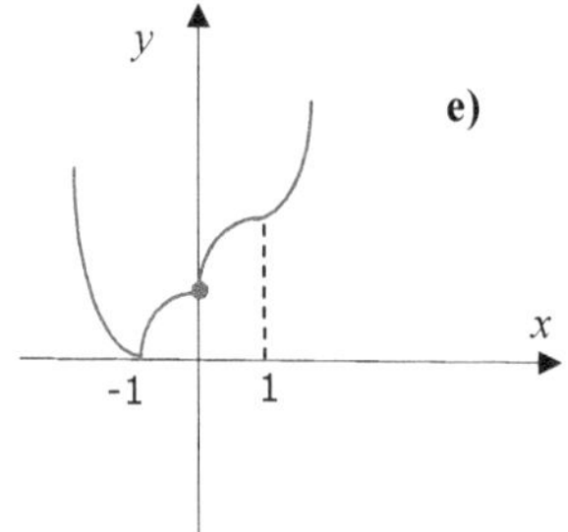

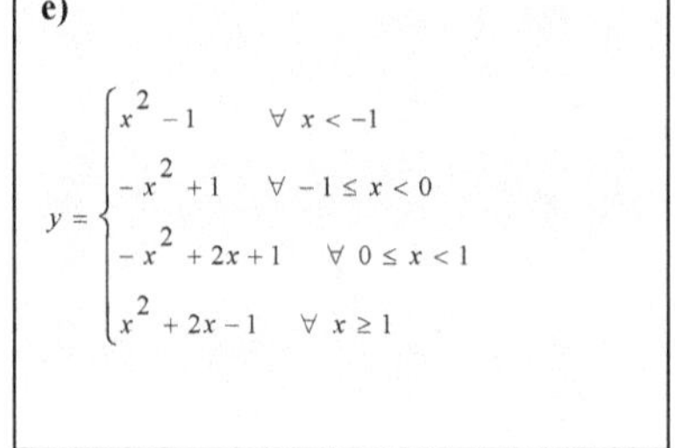

e)

$$y = \begin{cases} x^2 - 1 & \forall\, x < -1 \\ -x^2 + 1 & \forall\, -1 \le x < 0 \\ -x^2 + 2x + 1 & \forall\, 0 \le x < 1 \\ x^2 + 2x - 1 & \forall\, x \ge 1 \end{cases}$$

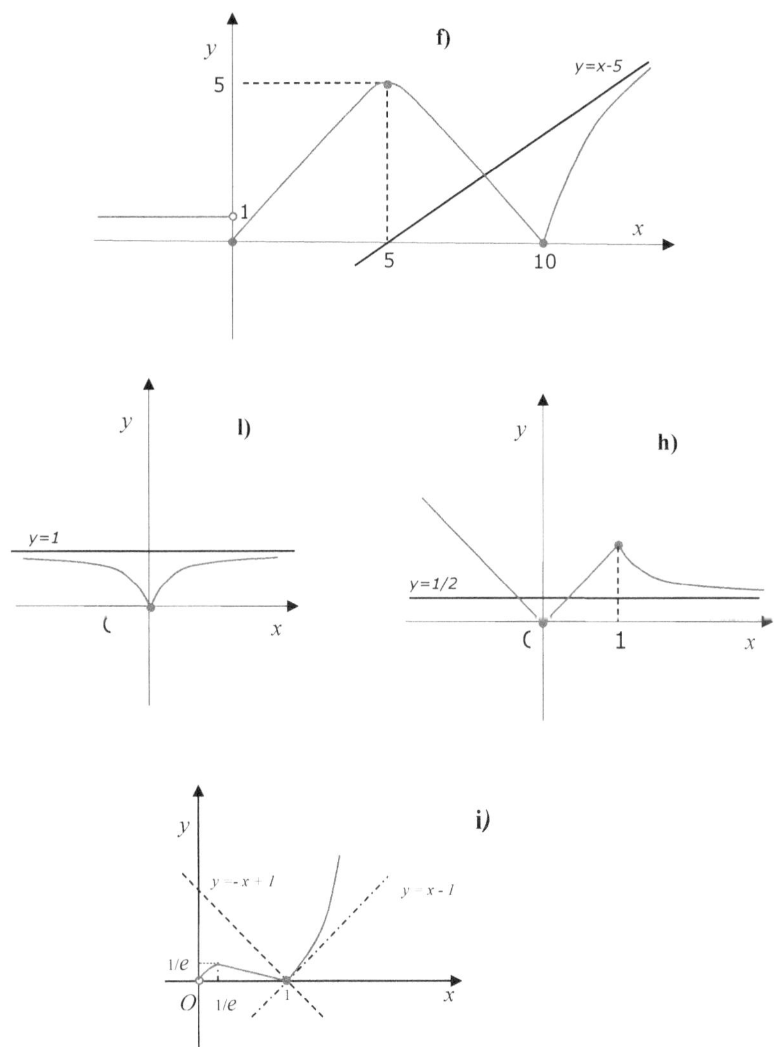

f)
y
x
5
5
10
y=x-5
1
l)
y
x
y=1
(
h)
y
x
y=1/2
(
1
i)
y
x
y=-x+1
y=x-1
1/e
O
1/e
1

7. Graph of other functions

N.1.- Graph the function $y = \dfrac{1}{4}\ln^2 x + 1^{\frac{x+1}{}}$.

DOMAIN D

$$\begin{cases} \ln^2 x + 1 \neq 0 \\ x > 0 \end{cases} \Rightarrow x > 0 \Rightarrow D =]0,+\infty[$$

SIGN OF FUNCTION

$$y \geq 0 \quad \Rightarrow \quad \frac{1}{4}\left(\frac{x+1}{\ln^2 x+1}\right) \geq 0 \quad \Rightarrow \quad \forall x \in D$$

LIMITS AND ASYMPTOTES

$$\lim_{x \to 0}\frac{1}{4}\ln^2 x+1^{\frac{x+1}{}} = \frac{1}{4}\lim_{x \to 0}\ln^2 x+1^{\frac{x+1}{}} = \frac{1}{4}^{\frac{1}{+\infty}} = \frac{1}{4}^0 = 1 \Rightarrow x = 1 \ \text{ vertical asymptote}$$

$$\lim_{x \to +\infty}\frac{1}{4}\log^2 x+1^{\frac{x+1}{}} \Rightarrow \lim_{x \to +\infty}\frac{x+1}{\log^2 x+1} = \frac{+\infty}{+\infty} = \lim_{x \to +\infty}\frac{x}{2} = +\infty \Rightarrow \lim_{t \to +\infty}\frac{1}{4}^t = 0 \Rightarrow$$

$$\Rightarrow y = 0 \ \text{ horizontal asymptote}$$

THE FIRST ORDER DERIVATIVE

$$y' = \frac{1}{4}^{\frac{x+1}{\log^2 x+1}} \cdot \log 1/4 \cdot \frac{\log^2 x + 1 - \dfrac{2(x+1)\log x}{x}}{\left(\log^2 x + 1\right)^2} \ ,$$

$$y' \geq 0 \quad \Rightarrow \quad \frac{1}{4}^{\frac{x+1}{\log^2 x+1}} \cdot \log 1/4 \cdot \frac{\log^2 x + 1 - \dfrac{2(x+1)\log x}{x}}{\left(\log^2 x + 1\right)^2} \geq 0 \quad \Rightarrow$$

$$\Rightarrow \ \log^2 x + 1 - \frac{2(x+1)\log x}{x} \leq 0 \quad \Rightarrow \quad \log^2 x + 1 \leq \frac{2(x+1)\log x}{x}$$

($\log 1/4 < 0$).

$$*)\ \log^2 x + 1 \le \frac{2(x+1)\log x}{x}$$

$$\downarrow$$

$$u(x) = \log^2 x + 1\,; \qquad v(x) = \frac{2(x+1)\log x}{x}\,.\ \ (\text{Figure 7.1.1})$$

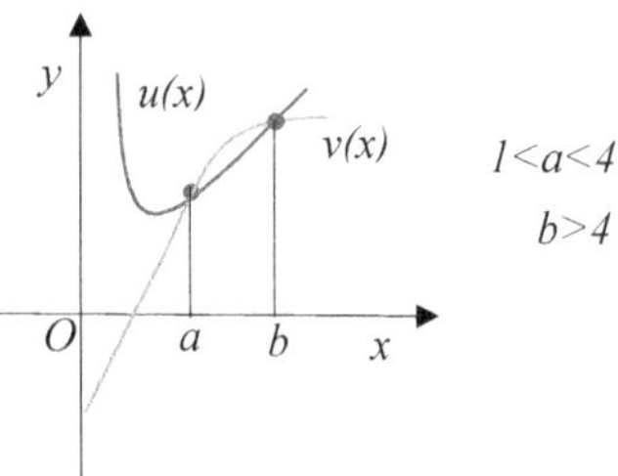

Figure 7.1.1

$$\downarrow$$

x	$u(x)$	$v(x)$
$x=4$	2,92	3,4

$$\downarrow$$

Local minimums and maximums: $N(a, y(a))$, $M(b, (y(b))$

Graph of the function *u(x)*

- Domain D
- $u(x) > 0 \ \forall x \in D$
- $\displaystyle \lim_{x \to 0} \log^2 x + 1 = +\infty \ \Rightarrow \ x = 0$ vertical asymptotes

$$\lim_{x \to +\infty} \log^2 x + 1 = +\infty \quad \text{e} \quad m = \lim_{x \to 0} \frac{1}{x}(\log^2 x + 1) = 0.$$

- $u' = \dfrac{2\log x}{x}; \ u' \geq 0 \Rightarrow \dfrac{2\log x}{x} \geq 0 \Rightarrow x \geq 1 \Rightarrow$ local minimum $N\,(1,1)$
- The graph of function *u(x)* is represented in Figure 7.1.1

Graph the function *v(x)*

- Domain D
- $v(x) > 0 \ \forall x>1, \ v(x) = 0 \ \forall x=1, \ v(x) < 0 \ \forall x: 0 < x < 1$
- $\displaystyle \lim_{x \to 0} \frac{2(x+1)\ln x}{x} = -\infty \ \Rightarrow \ x = 0$ vertical asymptotes

$$\lim_{x \to +\infty} \frac{2(x+1)\ln x}{x} = +\infty,$$

$$m = \lim_{x \to +\infty} \frac{2(x+1)\ln x}{x^2} = 0 \cdot (-\infty) \overset{DH}{=} \lim_{x \to +\infty} \frac{2x^2 + 4x}{x(2x+2)^2} = 0.$$

- $v' = \dfrac{1}{x^2}(-\ln x + 2x + 2); \ v' \geq 0 \ \Rightarrow \ \dfrac{1}{x^2}(-\ln x + 2x + 2) \geq 0 \Rightarrow \forall x \in D$

$$\downarrow$$

$$-\ln x + 2x + 2 \geq 0 \quad \Rightarrow \quad 2x + 2 \geq \ln x$$

$$\downarrow$$

$$h(x) = 2x + 2 ; \quad k(x) = \ln x \ (\text{Figure 7.1.1}).$$

- The graph of function *v(x)* is represented in Figure 7.1

The graph of function y is represented in Figure .7.17

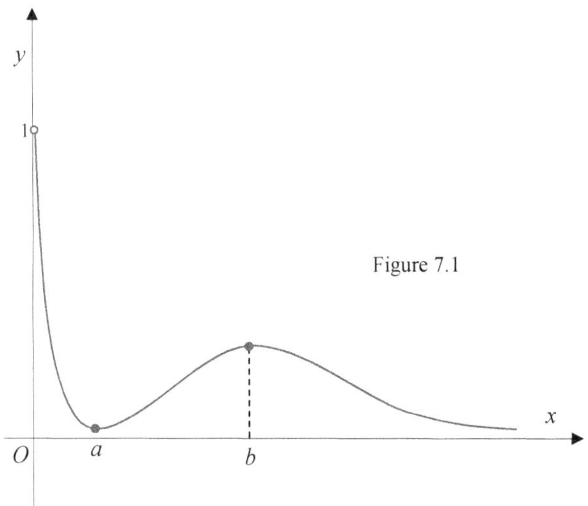

Figure 7.1

$A(1; 0,0607)$,
$B(2; 0,0625)$,
$C(4; 0,09)$,
$D(5; 0,22)$,
$E(6; 0,1)$,
$F(10; 0,08)$,
$G(100; 0,001748)$.

N.2.- Graph the function $y = \dfrac{x^2}{|x|} \ln|x|$

DOMAIN D

$$\begin{cases} |x| > 0 \\ |x| \neq 0 \end{cases} \Rightarrow |x| \neq 0 \Rightarrow x \neq 0 \Rightarrow D = R - \{0\}$$

$$y = y_1 \cup y_2$$

$$y_1 = \frac{x^2}{x} \ln x = x \ln x \quad \forall x > 0, \qquad y_2 = \frac{x^2}{-x} \ln(-x) = -x \ln(-x) \quad \forall x < 0$$

The graph of function $y = y_1 \cup y_2$ is represented in Figure 7.2 .

Graph the function y_1

- $D_1 =]0,+\infty[$
- $y_1 \geq 0 \quad \Rightarrow \quad x \ln x \geq 0 \quad \Rightarrow \quad \ln x \geq 0 \quad \Rightarrow \quad x \geq 1$
- $\lim\limits_{x \to 0} y_1 = \lim\limits_{x \to 0} x \ln x = 0, \qquad \lim\limits_{x \to +\infty} y_1 = \lim\limits_{x \to +\infty} x \ln x = +\infty$

$$m = \lim\limits_{x \to +\infty} \frac{y_1}{x} = \lim\limits_{x \to +\infty} \ln x = +\infty$$

- $y_1' = \ln x + 1$

  $y_1' \geq 0 \quad \Rightarrow \quad \ln x + 1 \geq 0 \quad \Rightarrow \quad \ln x \geq -1 \quad \Rightarrow \quad \ln x \geq \ln e^{-1} \quad \Rightarrow \quad x \geq e^{-1}$

  Local minimum: $x = e^{-1}$.
- The graph of function is represented in Figure 7.2.1.

Graph the function y_2

- $D_2 =]-\infty,0[$

- $y_2 \geq 0 \quad \Rightarrow \quad -x \ln(-x) \geq 0 \quad \Rightarrow \quad \ln(-x) \geq 0 \quad \Rightarrow \quad -x \geq 1 \Rightarrow \quad x \leq -1$

- $\lim\limits_{x \to 0} y_2 = \lim\limits_{x \to 0} -x \ln(-x) = 0, \quad \lim\limits_{x \to -\infty} y_2 = \lim\limits_{x \to -\infty} -x \ln(-x) = +\infty$

$$m = \lim\limits_{x \to -\infty} \frac{y_2}{x} = \lim\limits_{x \to -\infty} -\ln(-x) = -\infty$$

- $y_2' = -\ln(-x) - 1 = -[\ln(-x) + 1]$

  $y_2' \geq 0 \quad \Rightarrow \quad -[\ln(-x) + 1] \geq 0 \quad \Rightarrow \quad \ln(-x) \leq -1 \quad \Rightarrow \quad -x \leq e^{-1} \Rightarrow \quad x \geq -e^{-1}$

  Local minimum: $x = -e^{-1}$.

- The graph of function is represented in Figure 7.2.2

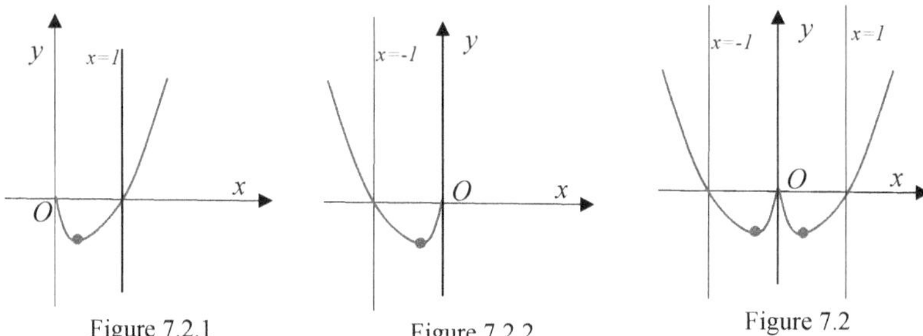

Figure 7.2.1   Figure 7.2.2   Figure 7.2

N.3. Graph the function $y = \ln(\sin x + \cos x)$.

SYMMETRIES AND PERIODICITIES

$$y(x + 2\pi) = \ln[\sin(x + 2\pi) + \cos(x + 2\pi)] = \ln(\sin x + \cos x) = y(x) \quad \Rightarrow$$

$$\Rightarrow y \text{ is a periodic function, period } p = 2\pi$$

DOMAIN D

$$\cos x + \operatorname{sen} x > 0 \ \Rightarrow \quad \cos x \,(\operatorname{tg} x + 1) > 0 \quad \Rightarrow \quad 0 \le x < 3/4\pi, \quad 7/4\pi < x \le \ 2\pi \ \Rightarrow$$

$$\Rightarrow D = \ [0 \ , 3/4\pi[\ \cup \ ]7/4\pi, 2\pi].$$

SIGN OF FUNCTION

$$y \ge 0 \ \Rightarrow \ \ln(\sin x + \cos x) \ge 0 \ \Rightarrow \ \sin x + \cos x \ge 1 \ \Rightarrow \ \frac{2t}{1+t^2} + \frac{1-t^2}{1+t^2} - 1 \ge 0 \ \Rightarrow$$

$$\Rightarrow \ 2t + 1 - t^2 - 1 - t^2 \ge 0 \ \Rightarrow \ -t^2 + t \ge 0 \Rightarrow \ 0 \le t \le 1 \ \Rightarrow \begin{cases} \tan\frac{x}{2} \le 1 \\ \tan\frac{x}{2} \ge 0 \end{cases}, \ \ (t = \tan\frac{x}{2}) \Rightarrow$$

$$\Rightarrow \ 0 \le x \ \le \pi/$$

$$\downarrow$$

$$0 < x \ < \pi/2 \ \Rightarrow \ y > 0$$
$$x = 0, x = \pi/2, x = 2\pi \ \Rightarrow \ y = 0$$

LIMITS AND ASYMPTOTES

$$\lim_{x \to \frac{3}{4}\pi} y = \lim_{x \to \frac{3\pi}{4}} \ln(\sin x + \cos x) = -\infty \ , \ \Rightarrow \ \ x = 3/4\pi \ \text{vertical asymptote}$$

$$\lim_{x \to \frac{7}{4}\pi} y = -\infty \ \Rightarrow \ \ x = 7/4\pi \ \text{vertical asymptote}$$

THE FIRST ORDER DERIVATIVE

$$y' = \frac{1}{\sin x + \cos x} \cdot (\cos x - \sin x) = \frac{\cos x - \sin x}{\sin x + \cos x}$$

$$y' \ge 0 \ \Rightarrow \ \frac{\cos x - \sin x}{\sin x + \cos x} \ge 0 \ \Rightarrow \ 0 < x < \pi/4 \ \vee \ 7/4\pi < x < 2\pi.$$

Local minimum: $N\left(\pi/4, \dfrac{1}{2}\log 2\right)$.

$$y\left(\frac{\pi}{4}\right) = \ln\left[\sin\left(\frac{\pi}{4}\right) + \cos\left(\frac{\pi}{4}\right)\right] = \ln\left(\frac{\sqrt{2}}{2} + \frac{\sqrt{2}}{2}\right) = \ln\left(\sqrt{2}\right) = \frac{1}{2}\log 2$$

The graph of function is represented in Figure 7.3

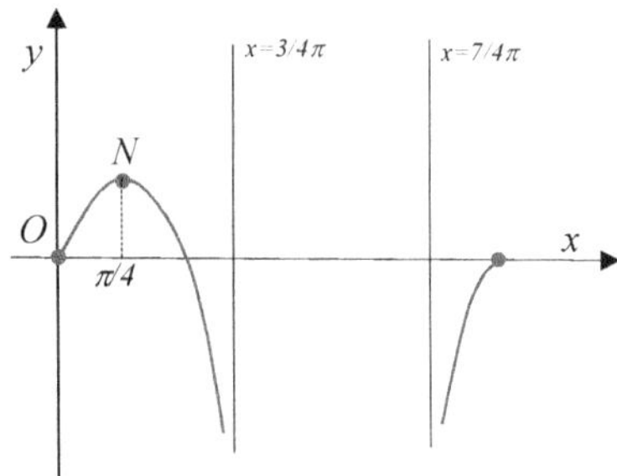

Figure 7.3

N.4. Graph the function $y = \left(\dfrac{|x-1|+3}{\sqrt{x}}\right)^{\pi}$.

DOMAIN D

$$\begin{cases} \dfrac{|x-1|+3}{\sqrt{x}} \geq 0 \\ x > 0 \end{cases} \Rightarrow \quad x > 0 \quad \Rightarrow \quad D = \,]\,0, +\infty\,[$$

SIGN OF FUNCTION

$$y \geq 0 \quad \Rightarrow \quad \left(\dfrac{|x-1|+3}{\sqrt{x}}\right)^{\pi} \geq 0 \quad \Rightarrow \forall x \in D$$

LIMITS AND ASYMPTOTES

$$\lim_{x \to 0}\left(\dfrac{|x-1|+3}{\sqrt{x}}\right)^{\pi} = +\infty \quad \Rightarrow \quad x = 0 \ \text{ vertical asymptote}$$

$$\lim_{x \to +\infty}\left(\dfrac{|x-1|+3}{\sqrt{x}}\right)^{\pi} = +\infty \, ,$$

$$m = \lim_{x \to +\infty} \dfrac{1}{x} \cdot \left(\dfrac{|x-1|+3}{\sqrt{x}}\right)^{\pi} = 0$$

THE FIRST ORDER DERIVATIVE

$$y' = \pi \left(\dfrac{x+2}{\sqrt{x}}\right)^{\pi-1} \cdot \dfrac{2-x}{2x\sqrt{x}} \qquad \forall \ x > 1 , \qquad\qquad y' = \pi \left(\dfrac{4-x}{\sqrt{x}}\right)^{\pi-1} \cdot \dfrac{-4-x}{2x\sqrt{x}} \qquad \forall \ 0 < x < 1$$

$$y' \geq 0 \Rightarrow \pi \left(\dfrac{x+2}{\sqrt{x}}\right)^{\pi-1} \cdot \dfrac{x-2}{2x\sqrt{x}} \geq 0 \quad \Rightarrow \quad x \geq 2, \qquad \forall \ x > 1$$

$$y' \geq 0 \Rightarrow \pi \left(\dfrac{4-x}{\sqrt{x}}\right)^{\pi-1} \cdot \dfrac{-4-x}{2x\sqrt{x}} \geq 0 \quad \Rightarrow \quad -4-x \geq 0 \quad \Rightarrow \quad x \leq -4 \qquad \forall \ 0 < x < 1$$

Local minimum: $N\left[2,\left(2\sqrt{2}\right)^{\pi}\right]$

$$y(2)=\left(\frac{|2-1|+3}{\sqrt{2}}\right)^{\pi}=\left(\frac{4}{\sqrt{2}}\right)^{\pi}=\left(2\sqrt{2}\right)^{\pi}$$

The graph of function is represented in Figure 7.4

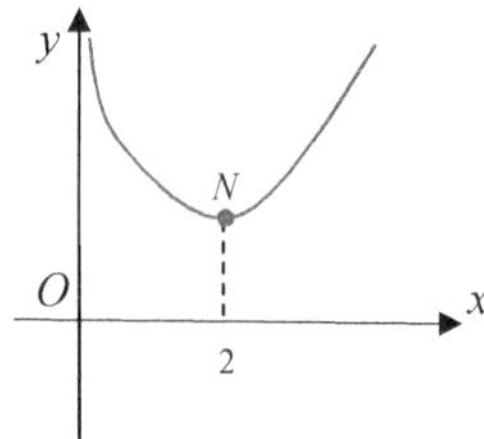

Figure 7.4

N.5.- Graph the function $y = \max\left\{x^2 - 3,\, 2x\right\}$.

<u>DOMAIN D</u>

$$D = R$$

$$y = \max\left\{x^2 - 3,\, 2x\right\} \;\Leftrightarrow\; \begin{cases} x^2 - 3 & \forall x : x^2 - 3 \geq 2x \\ 2x & \forall x : x^2 - 3 < 2x \end{cases} \;\Rightarrow$$

$$\downarrow$$

$$\Rightarrow\quad y = \begin{cases} x^2 - 3 & \forall x : \ x \leq -1 \vee x \geq 2 \\ 2x & \forall x : \ -1 < x < 2 \end{cases} \quad \text{(Figure 7.5.1)}$$

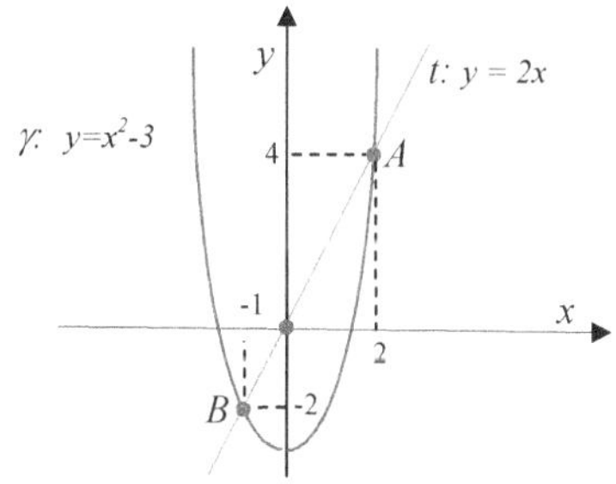

Figure 7.5.1

The graph of function $y = \max\left\{x^2 - 3,\, 2x\right\}$ is represented in fig 7.5

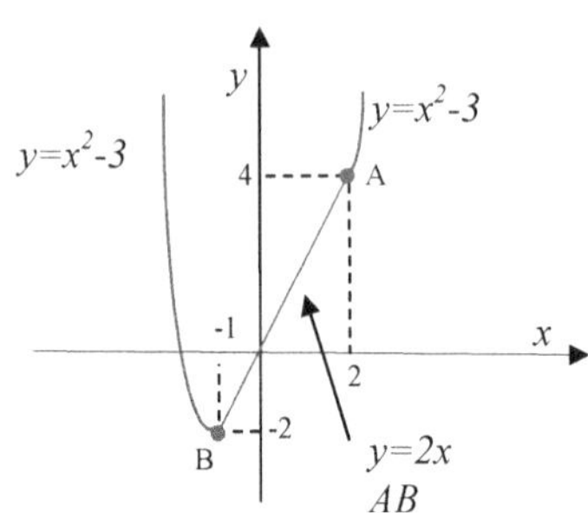

Figure 7.5

N.6.- Graph the function $y = \min\left\{\sqrt{2x-1}, x\right\}$.

<u>**DOMAIN D**</u>

$$2x+1 \geq 0 \quad \Rightarrow \quad x \geq -\frac{1}{2} \quad \Rightarrow D = \left[\frac{1}{2}, +\infty\right[\ .$$

Graph the function $y = \min\left\{\sqrt{2x-1}, x\right\} \ \forall x \in D$

$$y = \min\left\{\sqrt{2x-1}, x\right\} \quad \Leftrightarrow \quad y = \begin{cases} x & \forall x : \dfrac{1}{2} \leq x < 1 \\[2mm] \sqrt{2x-1} & \forall x : x \geq 1 \end{cases}$$

The graph of function $y = \min\left\{\sqrt{2x-1}, x\right\}$ is represented in fig 7.6

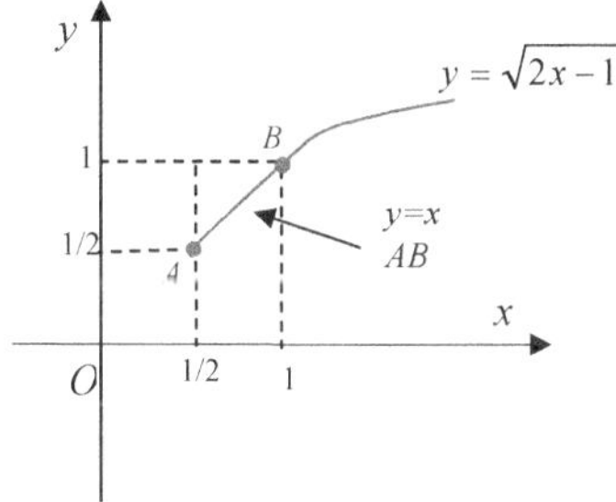

Figure 7.6

N.7.- Graph the function $y = \sqrt{x}^{\sqrt{x}}$.

DOMAIN D

$$\begin{cases} \sqrt{x} > 0 \\ x \geq 0 \end{cases} \Rightarrow x > 0. \Rightarrow D = \,]0. +\infty[$$

$$y = f(x)^{g(x)} = e^{g(x)\log f(x)} \Rightarrow y = \sqrt{x}^{\sqrt{x}} = e^{\sqrt{x}\,\ln\sqrt{x}}$$

SIGN OF FUNCTION

$$y \geq 0 \Rightarrow \sqrt{x}^{\sqrt{x}} \geq 0 \Rightarrow e^{\sqrt{x}\ln\sqrt{x}} \geq 0 \Rightarrow \forall x \in D$$

LIMITS AND ASYMPTOTES

$$\lim_{x\to 0} y = \lim_{x\to 0} \sqrt{x}^{\sqrt{x}} = \lim_{x\to 0} e^{\sqrt{x}\ln\sqrt{x}} \Rightarrow \lim_{x\to 0} \sqrt{x}\ln\sqrt{x} = 0(-\infty) \Rightarrow$$

$$\Rightarrow \lim_{x\to 0} \frac{\ln\sqrt{x}}{\frac{1}{\sqrt{x}}} = \frac{-\infty}{+\infty} \Rightarrow$$

$$\Rightarrow \lim_{x\to 0} \frac{\ln\sqrt{x}}{\frac{1}{\sqrt{x}}} \overset{DH}{=} \lim_{x\to 0} \frac{\frac{1}{\sqrt{x}}\frac{1}{2}(x)^{-\frac{1}{2}}}{-\frac{1}{2x\sqrt{x}}} = \lim_{x\to 0} \frac{\frac{1}{2x}}{\frac{-1}{2x\sqrt{x}}} = -\lim_{x\to 0}\sqrt{x} = 0 \Rightarrow$$

$$\Rightarrow \lim_{x\to 0} \sqrt{x}^{\sqrt{x}} = \lim_{x\to 0} e^{\sqrt{x}\ln\sqrt{x}} = \lim_{t\to 0} e^{t} = 1 \Rightarrow$$

$$\Downarrow$$
$$\Rightarrow P(0,1) \text{ removable discontinuity}$$

$$\lim_{x\to +\infty} \sqrt{x}^{\sqrt{x}} = \lim_{x\to +\infty} e^{\sqrt{x}\ln\sqrt{x}} = +\infty$$

$$m = \lim_{x\to +\infty} \frac{f(x)}{x} = \lim_{x\to +\infty} \frac{\sqrt{x}^{\sqrt{x}}}{x} = \lim_{x\to +\infty} \frac{e^{\sqrt{x}\ln x}}{x} = +\infty$$

THE FIRST ORDER DERIVATIVE

$$y' = e^{\sqrt{x}\ln\sqrt{x}}\left[\frac{1}{2}x^{-\frac{1}{2}}\ln\sqrt{x} + \frac{1}{2\sqrt{x}}\right] = e^{\sqrt{x}\ln\sqrt{x}}\left[\frac{\ln\sqrt{x}+1}{2\sqrt{x}}\right]$$

$$y' \geq 0 \;\Rightarrow\; e^{\sqrt{x}\ln\sqrt{x}}\left[\frac{\ln\sqrt{x}+1}{2\sqrt{x}}\right] \geq 0 \;\Rightarrow\; \frac{\ln\sqrt{x}+1}{2\sqrt{x}} \geq 0 \;\Rightarrow\; \ln\sqrt{x}+1 \geq 0 \;\Rightarrow$$

$$\Rightarrow\; x \geq \frac{1}{e^2}.$$

Local minimum: $N\left(\dfrac{1}{e^2}; \dfrac{1}{e}^{\frac{1}{e}}\right)$.

The graph of function is represented in Figure 7.7.

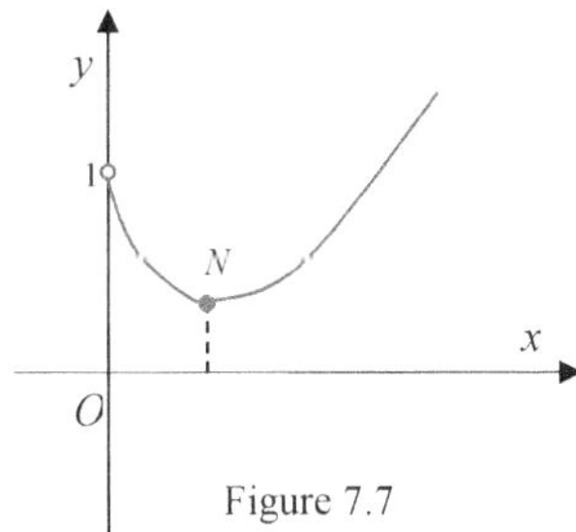

Figure 7.7

N.8.- Graph the function $y = x^{\left(\frac{1}{x}-1\right)}$.

DOMAIN *D*

$$\begin{cases} x > 0 \\ x \neq 0 \end{cases} \Rightarrow x > 0 \Rightarrow D = \left]0,+\infty\right[$$

SIGN OF FUNCTION

$$y \geq 0 \quad \Rightarrow \quad x^{\left(\frac{1}{x}-1\right)} \geq 0 \quad \Rightarrow \quad \forall x \in D$$

LIMITS AND ASYMPTOTES

$$\lim_{x \to 0} x^{\left(\frac{1}{x}-1\right)} = \lim_{x \to 0} e^{\left(\frac{1}{x}-1\right)\log x} \Rightarrow \lim_{x \to 0}\left(\frac{1}{x}-1\right)\log x = -\infty \Rightarrow$$

$$\Rightarrow \lim_{x \to 0} x^{\left(\frac{1}{x}-1\right)} = \lim_{x \to 0} e^{\left(\frac{1}{x}-1\right)\log x} = \lim_{t \to -\infty} e^{t} = 0^{+} \Rightarrow$$

$$\Downarrow$$

$$\Rightarrow \quad P(0,0) \text{ removable discontinuity}$$

$$\lim_{x \to +\infty} x^{\left(\frac{1}{x}-1\right)} = \lim_{x \to +\infty} e^{\left(\frac{1}{x}-1\right)\log x} = \lim_{x \to +\infty} e^{\left(\frac{1-x}{x}\right)\log x} = e^{(-1)(+\infty)} = 0 \Rightarrow$$

$$\Rightarrow \quad y = 0 \text{ horizontal asymptote}$$

THE FIRST ORDER DERIVATIVE

$$y' = e^{\left(\frac{1-x}{x}\right)\ln x}\left[-\frac{1}{x^2}\ln x + \left(\frac{1}{x}-1\right)\frac{1}{x}\right] = e^{\left(\frac{1-x}{x}\right)\ln x}\left[\frac{1}{x^2}(1-\ln x)-\frac{1}{x}\right]$$

$$y' \geq 0 \Rightarrow \frac{1}{x^2}(1-\ln x)-\frac{1}{x} \geq 0 \Rightarrow \frac{1}{x}\left[\frac{1}{x}(1-\ln x)-1\right] \geq 0 \Rightarrow \frac{1}{x}(1-\ln x)-1 \geq 0 \Rightarrow$$

$$\Rightarrow \frac{1}{x}-\frac{\ln x}{x}-1 \geq 0 \quad \Rightarrow \quad \frac{1-\ln x - x}{x} \geq 0 \quad \Rightarrow$$

$$\Rightarrow \quad 1 - \ln x - x \geq 0 \quad \Rightarrow \quad 1 - x \geq \ln x \quad \Rightarrow \quad 0 < x \leq 1$$

$$\downarrow$$

$$f(x) = 1 - x \; ; \quad g(x) = \ln x \quad \text{(Figure 7.8.1)}$$

$$\downarrow$$

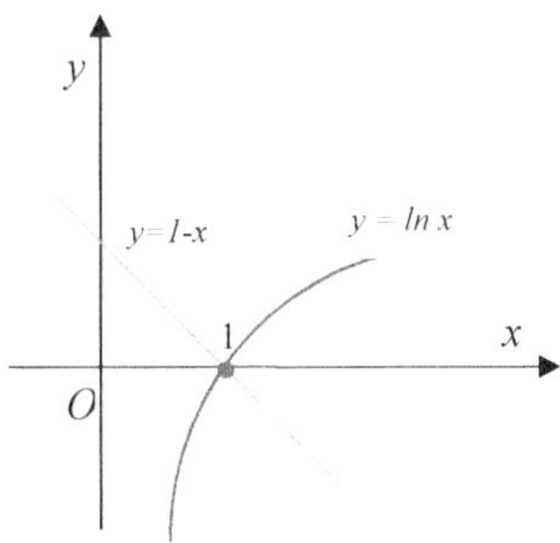

Figure 7.8.1

Local maximum: $M(1,1)$

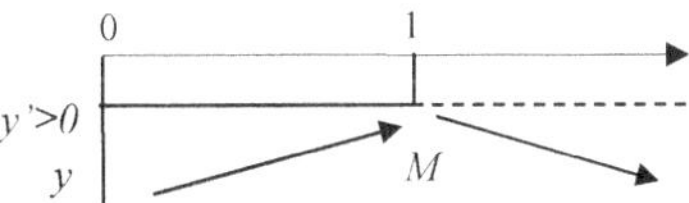

The graph of function is represented in Figure 7.8.

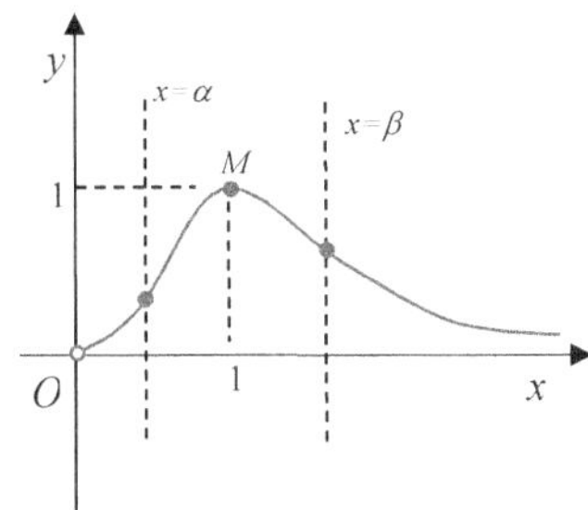

Figure 7.8

$$\lim_{x \to 0} y' = \lim_{x \to 0} e^{\left(\frac{1}{x}-1\right)\log x}\left(\frac{1-\log x - x}{x^2}\right) = \lim_{x \to 0} x^{\frac{1}{x}-3}(1-x-\log x) = 0$$

THE SECOND ORDER DERIVATIVE

$$y'' = \frac{e^{\left(\frac{1}{x}-1\right)\ln x}}{x^4}\left(-2x^2 + 5x - 1 - 4x\ln x - \ln^2 x + 2\ln x\right)$$

$$y'' = 0, \quad \Rightarrow \quad -2x^2 + 5x - 1 - 4x\ln x - \ln^2 x + 2x\ln x = 0 \quad \Rightarrow$$

$$\Rightarrow \quad -2x^2 + 5x - 1 - 4x\ln x = \ln^2 x - 2x\ln x$$

$$\downarrow$$

$$f(x) = -2x^2 + 5x - 1 - 4x\ln x, \qquad g(x) = \ln g^2 x - 2\ln x \text{ (Figure 7.8.2)}$$

Inflection points: $x = \alpha \in\ ]0,1[,\ x = \beta\ (\beta > 1)$

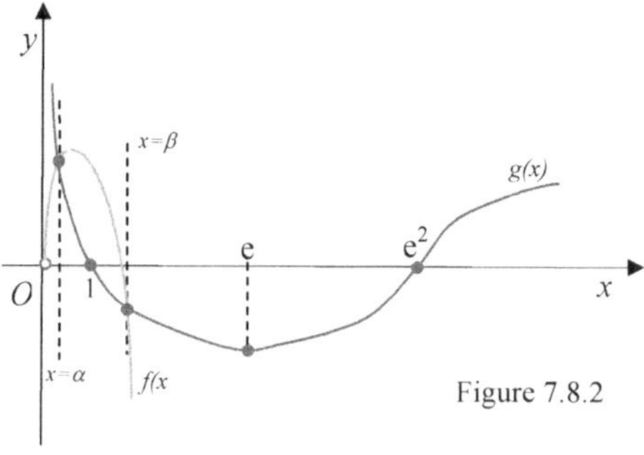

Figure 7.8.2

Table 1

x	$f(x)$	$g(x)$
e^{-1}	2,04	2,67
0,4	2,14	2,67
0,445	2,27018	2,27494
0,45	2,282	2,235
1	2	0
1,4	0,2	- 0,56
1,5	- 0,68	- 0,65
1,55	- 0,772	- 0,684

$$\alpha \approx 0{,}445\,5 \quad \Rightarrow \quad F_1\,(0{,}445\;;\;0{,}3543)$$
$$\beta \approx 1{,}5 \quad \Rightarrow \quad F_2\,(1{,}5\;;\;0{,}875)$$

Exercises.-

A) Graph the functions .

a) $\quad y = \left(x^2 - 1\right) e^{-\left(x^2\right)}$,

b) $\quad y = e^{2\sin\left(x+\frac{\pi}{3}\right)} \quad \forall x \in \left[-\frac{\pi}{2}, \frac{5}{3}\pi\right]$.

c) $\quad y = \begin{cases} \dfrac{e}{x^4} & \forall \ x < -1 \\ x^2 - 1 & \forall \ -1 \le x \le 1 \\ \dfrac{1}{x \cdot e^{\ln x}} & \forall \ x > 1 \end{cases}$,

d) $\quad y = \begin{cases} \dfrac{e^{-\frac{1}{x}}\left(x^2 + x + 2\right)}{x^2} & \forall \ x > 0 \\ -x^2 - x & \forall \ -1 \le x < 0 \\ 0 & \forall \ x < -1 \end{cases}$

e) $\quad y = \lim_{n \to \infty} \dfrac{x}{1 + x^n} \ , \quad \forall x \in [0,+\infty[$

e]

$y = \begin{cases} x & \forall 0 \le x < 1 \\ \dfrac{1}{2} & \forall x = 1 \\ 0 & \forall x > 1 \end{cases}$

f) $\quad y = \lim_{a \to 0} \sqrt{a^2 + x^2}$.

g]

$y = \begin{cases} 6 - 2x & \forall \ 0 \le x \le 3 \\ x^2 - 5x + 6 & \forall x < 0, \ x > 3 \end{cases}$

g) $\quad y = \max\left\{6 - 2x, \ x^2 - 5x + 6\right\}$

h) $\quad y = \min\left\{x, \ x^2\right\}$

h]

$y = \begin{cases} x^2 & \forall \ 0 \le x \le 1 \\ x & \forall x < 0, x > 1 \end{cases}$

i) $\quad y = \max\left\{\dfrac{1}{x}, \ x\right\}$

l) $\quad y = \begin{cases} \dfrac{e^{x-1}}{\sqrt[3]{x-1}} & \forall \ x \le 0 \\ 2x - 3 & \forall \ 0 < x \le 1, \\ \dfrac{-2x^2}{\ln x} & \forall \ x > 1 \end{cases}$

m) $\quad y = \begin{cases} \ln\dfrac{(x-2)^2}{x} & \forall \ x > 0 \\[3mm] e^{\frac{2x^2-5x}{x+3}} & \forall \ x \leq 0 \end{cases}$

n) $\quad y = \dfrac{2^{1-|x^2-1|}}{x-1}$,

o) $\quad y = \begin{cases} \dfrac{(2\ln x - 9)x}{\ln x} & \forall \ x > 0 \\[3mm] 8 & \forall \ x = 0 \\[3mm] e^{\frac{1}{x}}\dfrac{2x^2 - 2x + 1}{x^2} & \forall \ x < 0 \end{cases}$

p) $\quad y = \begin{cases} 2x-1-\ln\left(e^x + 2\right) & \forall \ x \leq 0 \\[3mm] \dfrac{(x^2 + x + 2)e^{\frac{1}{x}}}{x^2} & \forall \ x > 0 \end{cases}$,

q) $\quad y = \begin{cases} x - \arctan x + \dfrac{2}{3}\ln(1 + x^2) & \forall \ x < -1/2 \\[3mm] x^2 - 1 & \forall \ -1/2 \leq x < 1 \\[3mm] (x-1)e^{\frac{\pi}{4}-\arctan\frac{x-2}{x-1}} & \forall \ x \geq 1 \end{cases}$

r) $\quad y = \sin\left(\dfrac{1}{x}\right)$,

s) $y = x\sin\left(\dfrac{1}{x}\right)$,

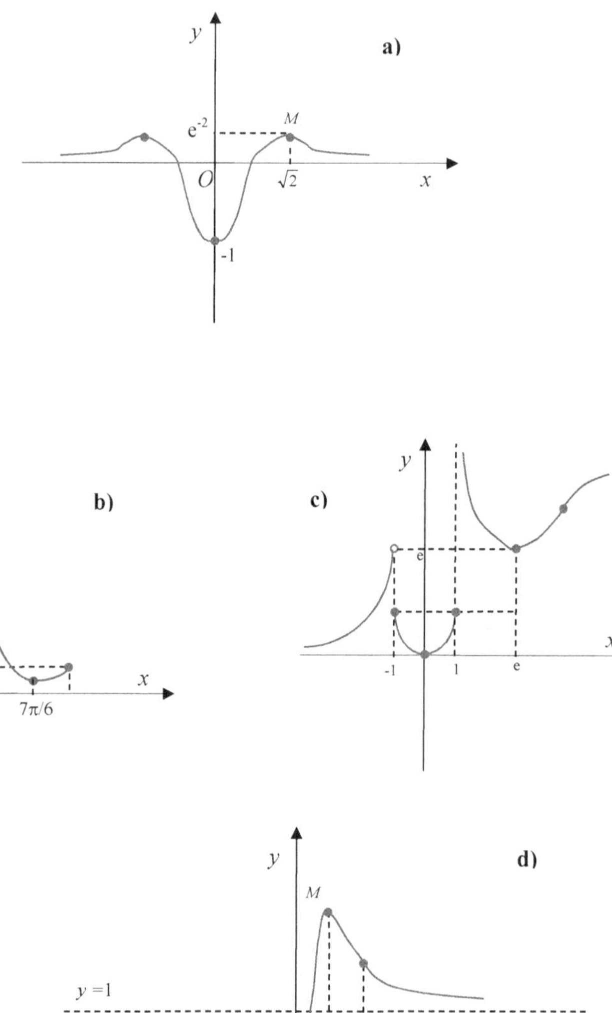

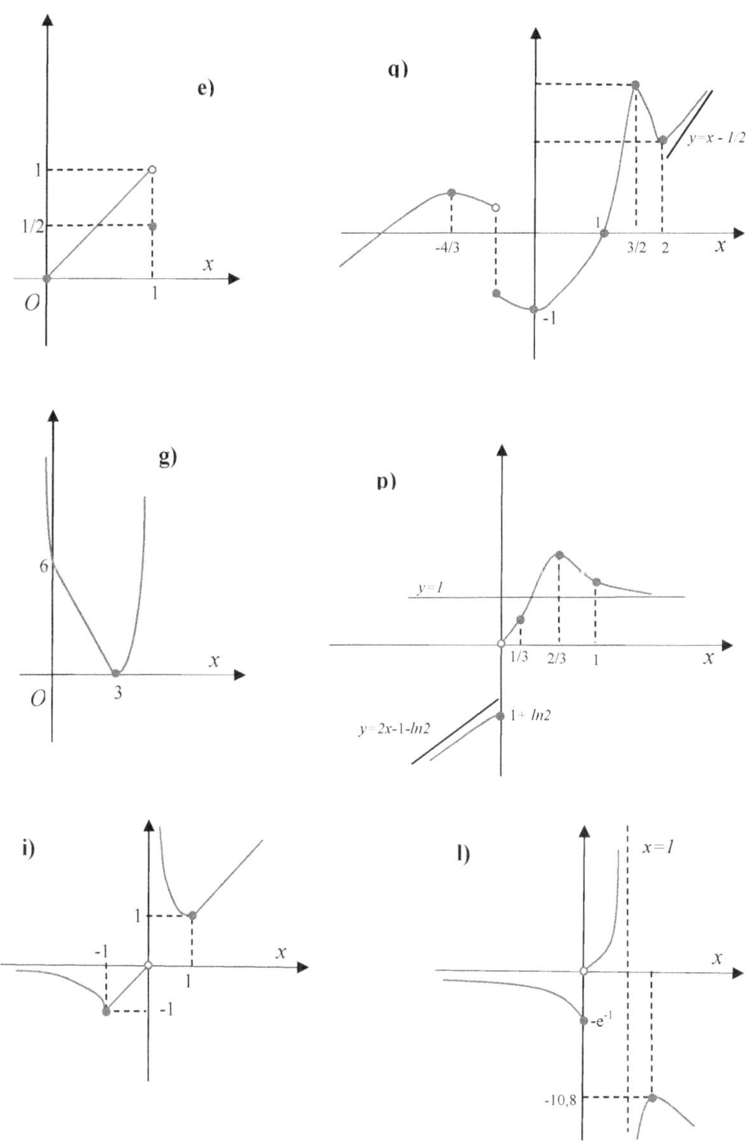
e)
q)
1
1/2
O
1
x
y=x - 1/2
-4/3
1
3/2
2
x
-1
g)
p)
6
O
3
x
y=1
y=2x-1-ln2
1+ ln2
1/3
2/3
1
x
i)
l)
x=1
1
-1
1
x
-e⁻¹
-1
-10,8
x

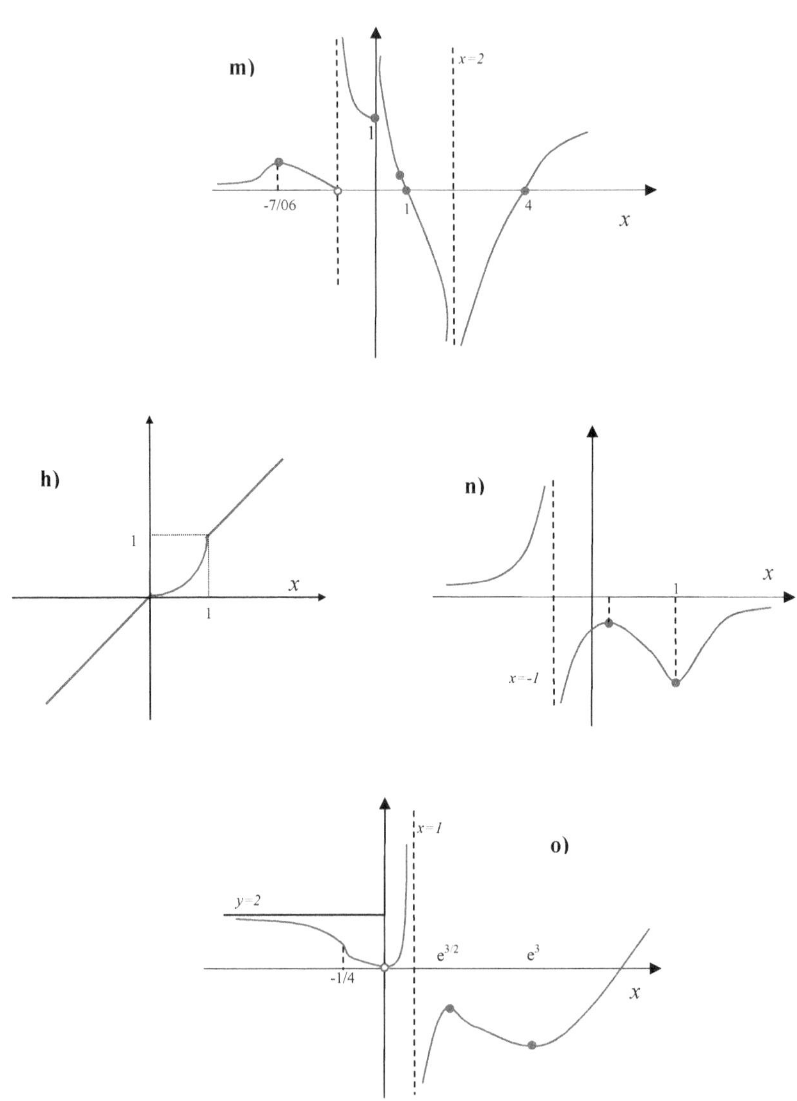
m)
x=2
1
-7/06
1
4
x
h)
1
1
x
n)
x=-1
1
x
o)
x=1
y=2
e^{3/2}
e^3
-1/4
x

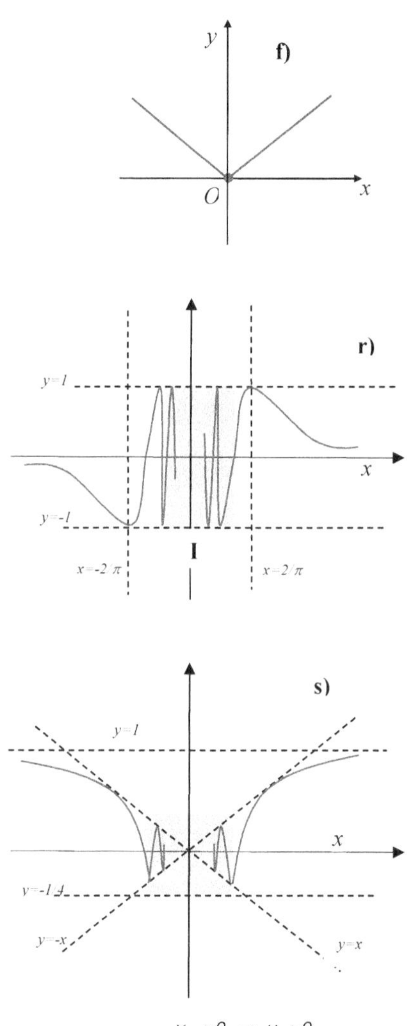

f)
y
x
O
r)
y=1
y=-1
x=-2/π
x=2/π
l
s)
y=1
y=-1
y=-x
y=x
x
x → 0 ⇒ y → 0

B) Graph the functions:

1) $y = x\sin x$, **2)** $y = e^{\sin x}$, **3)** $y = \sin(e^x)$

4) $y = -x + 1 + \dfrac{x}{2}\ln\dfrac{1+x^2}{4} + \arctan x$ **5)** $y = e + \ln x + \arctan\dfrac{1}{\ln x}$,

6) $y = \dfrac{3 + 2\sqrt{3}\tan x - 3\tan^2 x}{1 + \tan^2 x}$, **7)** $y = \sin x^{\sin x}$, **8)** $y = \dfrac{\left|1 + 3\ln x\right|}{|x|^3}$,

9) $y = x + 1 + 2\arcsin\dfrac{e^x - \sqrt{e}}{e^x + \sqrt{e}}$, **10)** $y = x + e^2 \arctan\sqrt{1-x}$.

11) $y = \dfrac{e^2 + \pi}{3\sqrt{10 - e}}x + \dfrac{1}{5 + e\pi}$; **12)** $y = \left(\dfrac{3e^{42} + \pi^6}{3\log 2 + \sqrt{1+e}}\right)^{\frac{7+\pi}{1+e}}$.

www.matematicus.com

APPENDIX

Algebra

1) $ax + b = 0 \quad \Rightarrow \quad x = -\dfrac{b}{a}, \ a \neq 0$

2) $ax^2 + bx + c = 0 \quad \Rightarrow \quad x_{1,2} = \dfrac{-b \pm \sqrt{b^2 - 4ac}}{2a}, \quad a \neq 0$

3) $\sqrt{f(x)} = g(x) \quad \Rightarrow \quad \begin{cases} f(x) = [g(x)]^2 \\ g(x) \geq 0 \end{cases}$

4) $\sqrt[3]{f(x)} = g(x) \quad \Rightarrow \quad f(x) = [g(x)]^3$

5) $a^x = b \quad \Rightarrow \quad x = \log_a b, \quad a > 0, a \neq 1, \quad b > 0$

6) $\log_a x = b \quad \Rightarrow \quad x = a^b, \quad a > 0, a \neq 1, b > 0$

7) $\sin x = m \quad \Rightarrow \quad x = \arcsin m, \ x = \pi - \arcsin m, \quad m \in [-1, 1]$

8) $\cos x = p \quad \Rightarrow \quad x = \arccos p, \ x = 2\pi - \arccos p, \ p \in [-1, 1]$

9) $\tan x = q \quad \Rightarrow \quad x = \arctan(q), \ x = \pi + \arctan(q)$

10) $\arcsin x = m \quad \Rightarrow \quad x = \sin m, \quad m \in [\, -\pi/2, \, \pi/2]$

11) $\arccos x = n \quad \Rightarrow \quad x = \cos n, \quad n \in [\, 0, \pi]$

12) $\arctan x = k \quad \Rightarrow \quad x = \tan k, \quad k \in \,]\, -\pi/2, \, \pi/2[$

13) $\text{arc} \cot x = h \quad \Rightarrow \quad x = \cot h, \quad h \in \,]\, 0, \pi\, [$

14) $|f(x)| = g(x) \quad \Rightarrow \quad \begin{cases} f(x) \geq 0 \\ f(x) = g(x) \end{cases} \cup \begin{cases} f(x) < 0 \\ -f(x) = g(x) \end{cases}$

15) $\quad ax+b>0 \quad \Rightarrow \quad \begin{cases} x>-\dfrac{b}{a} & a>0 \\[2ex] x<-\dfrac{b}{a} & a<0 \end{cases}$

16) $\quad ax+b<0 \quad \Rightarrow \quad \begin{cases} x<-\dfrac{b}{a} & a>0 \\[2ex] x>-\dfrac{b}{a} & a<0 \end{cases}$

17) $\quad ax^2+bx+c>0 \quad \Rightarrow \quad \begin{cases} x_1<x<x_2 & \Delta>0,\ a<0 \\[1ex] x<x_1, x>x_2 & \Delta>0,\ a>0 \\[1ex] \forall x \in R - \left\{-\dfrac{b}{2a}\right\} & \Delta=0,\ a>0 \\[1ex] \varnothing & \Delta=0,\ a<0 \\[1ex] \forall x \in R & \Delta<0,\ a>0 \\[1ex] \varnothing & \Delta<0,\ a<0 \end{cases}$

18) $\quad ax^2+bx+c<0 \quad \Rightarrow \quad \begin{cases} x_1<x<x_2 & \Delta>0,\ a>0 \\[1ex] x<x_1,\ x>x_2 & \Delta>0,\ a<0 \\[1ex] \forall x \in R - \left\{-\dfrac{b}{2a}\right\} & \Delta=0,\ a<0 \\[1ex] \varnothing & \Delta=0,\ a>0 \\[1ex] \forall x \in R & \Delta<0,\ a<0 \\[1ex] \varnothing & \Delta<0,\ a>0 \end{cases}$

19) $\quad \dfrac{f(x)}{g(x)}>0 \quad \Rightarrow \quad \begin{cases} f(x)>0 \\ g(x)>0 \end{cases} \cup \begin{cases} f(x)<0 \\ g(x)<0 \end{cases}$

20) $\quad \dfrac{f(x)}{g(x)}<0 \quad \Rightarrow \quad \begin{cases} f(x)>0 \\ g(x)<0 \end{cases} \cup \begin{cases} f(x)<0 \\ g(x)>0 \end{cases}$

21) $\quad f(x)g(x)>0 \quad \Rightarrow \quad \begin{cases} f(x)>0 \\ g(x)>0 \end{cases} \cup \begin{cases} f(x)<0 \\ g(x)<0 \end{cases}$

22) $\quad f(x)g(x)<0 \quad \Rightarrow \quad \begin{cases} f(x)>0 \\ g(x)<0 \end{cases} \cup \begin{cases} f(x)<0 \\ g(x)>0 \end{cases}$

23) $\quad \sqrt[n]{P(x)} > Q(x) \quad \Rightarrow \quad P(x) > \left[Q(x)\right]^n, \quad n \text{ odd}$

24) $\quad \sqrt[n]{P(x)} > Q(x) \quad \Rightarrow \quad \begin{cases} P(x) \geq 0 \\ Q(x) < 0 \end{cases} \cup \begin{cases} Q(x) \geq 0 \\ P(x) > [Q(x)]^n \end{cases} \quad , n \text{ even}$

25) $\quad \sqrt[n]{P(x)} < Q(x) \quad \Rightarrow \quad P(x) < \left[Q(x)\right]^n \quad , n \text{ odd}$

26) $\quad \sqrt[n]{P(x)} < Q(x) \quad \Rightarrow \quad \begin{cases} Q(x) > 0 \\ P(x) \geq 0 \\ P(x) < [Q(x)]^n \end{cases} \quad , n \text{ even}$

27) $\quad \left|f(x)\right| > g(x) \quad \Rightarrow \quad \begin{cases} f(x) \geq 0 \\ f(x) > g(x) \end{cases} \cup \begin{cases} f(x) < 0 \\ -f(x) > g(x) \end{cases}$

28) $\quad \left|f(x)\right| < g(x) \quad \Rightarrow \quad \begin{cases} f(x) \geq 0 \\ f(x) < g(x) \end{cases} \cup \begin{cases} f(x) < 0 \\ -f(x) < g(x) \end{cases}$

www.matematicus.com

Limits

1) $\quad \lim_{x \to c} [f(x) \pm g(x)] = \lim_{x \to c} f(x) \pm \lim_{x \to c} g(x)$

2) $\quad \lim_{x \to c} [f(x) \cdot g(x)] = \lim_{x \to c} f(x) \cdot \lim_{x \to c} g(x)$

3) $\quad \lim_{x \to c} \dfrac{f(x)}{g(x)} = \dfrac{\lim_{x \to c} f(x)}{\lim_{x \to c} g(x)}, \qquad \lim_{x \to c} g(x) \neq 0$

4) $\lim_{x \to c} \sqrt[n]{f(x)} = \sqrt[n]{\lim_{x \to c} f(x)}$

5). $\lim_{x \to c} k^{f(x)} = k^{\left[\lim_{x \to c} f(x) \right]}$

6) $\quad \lim_{x \to c} \log f(x) = \log \lim_{x \to c} f(x)$

7) $\begin{pmatrix} \lim_{x \to c} f(x) = l \\[4pt] f(x) \neq l \ \ \forall x \in I_c \\[4pt] \lim_{x \to l} g(x) = q \end{pmatrix} \Rightarrow \lim_{x \to c} g(f(x)) = q$

8) $\forall \, m \in R ,:$

$$m + (+\infty) = +\infty, \quad m + (-\infty) = -\infty, \quad (-\infty) + (-\infty) = -\infty, \quad (+\infty) + (+\infty) = +\infty.$$

$$(+\infty) \cdot (+\infty) = +\infty, \quad (-\infty) \cdot (-\infty) = +\infty, \quad (+\infty) \cdot (-\infty) = -\infty, \quad (-\infty) \cdot (+\infty) = -\infty.$$

$$m \cdot (+\infty) = +\infty \quad m > 0, \qquad m \cdot (+\infty) = -\infty \quad m < 0,$$

$$m \cdot (-\infty) = -\infty \quad m > 0, \qquad m \cdot (-\infty) = +\infty \quad m < 0.$$

$$\frac{m}{+\infty} = \begin{cases} 0^+ & \forall m > 0 \\ 0^- & \forall m < 0 \end{cases}, \quad \frac{m}{-\infty} = \begin{cases} 0^+ & \forall m < 0 \\ 0^- & \forall m > 0 \end{cases}, \quad \frac{m}{0^+} = \begin{cases} +\infty & \forall m > 0 \\ -\infty & \forall m < 0 \end{cases}, \quad \frac{m}{0^-} = \begin{cases} +\infty & \forall m < 0 \\ -\infty & \forall m > 0 \end{cases},$$

$$\frac{+\infty}{0^+}=+\infty,\quad \frac{0^+}{+\infty}=0^+,\quad \frac{+\infty}{0^-}=-\infty,\quad \frac{0^-}{+\infty}=0^-,\quad \frac{-\infty}{0^+}=-\infty,\quad \frac{0^+}{-\infty}=0^-,\quad \frac{-\infty}{0^-}=+\infty,\quad \frac{0^-}{-\infty}=0^+$$

$$m^{+\infty}=\begin{cases}+\infty & \forall\, m>1\\[4pt] 0 & \forall m:\,0\le m<1\end{cases}\qquad m^{-\infty}=\begin{cases}+\infty & \forall\, m:\,0\le m<1\\[4pt] 0 & \forall\, m>1\end{cases},$$

$$0^m=\begin{cases}+\infty & m<0\\[4pt] 0 & m>0\end{cases},\qquad (+\infty)^m=\begin{cases}+\infty & m>0\\[4pt] 0 & m<0\end{cases}.$$

9) Indeterminate forms:

$$\frac{0}{0}\;,\;\;\frac{\pm\infty}{\pm\infty}\;,\;+\infty-\infty,\;-\infty+\infty,\;,\;0(\pm\infty)\;,\;1^{\pm\infty},\;0^0,\;(+\infty)^0$$

10)
$$\lim_{x\to+\infty}\frac{a_0+a_1x+a_2x^2+....+a_nx^n}{b_0+b_1x+b_2x^2+...+b_mx^m}=\begin{cases}0 & n<m\\[8pt] \dfrac{a_n}{b_m}\cdot(+\infty) & n>m\;,\\[8pt] \dfrac{a_n}{b_m} & n=m\end{cases}$$

11)
$$\lim_{x\to-\infty}\frac{a_0+a_1x+a_2x^2+....+a_nx^n}{b_0+b_1x+b_2x^2+...+b_mx^m}=\begin{cases}0 & n<m\\[8pt] \dfrac{a_n}{b_m}\cdot(\pm\infty) & n>m\;\begin{cases}+\infty & n-m\ \text{even}\\[4pt] -\infty & n-m\ \text{odd}\end{cases}\\[12pt] \dfrac{a_n}{b_m} & n=m\end{cases}$$

12) L'Hôpital's rule (Bernoulli's rule)

$$\left. \begin{array}{l} \lim_{x \to c} f(x) = 0 \\ \lim_{x \to c} g(x) = 0 \end{array} \right) \quad \Rightarrow \quad \lim_{x \to c}\frac{f(x)}{g(x)} = \lim_{x \to c}\frac{f'(x)}{g'(x)}$$

$$\left. \begin{array}{l} \lim_{x \to c} f(x) = \pm\infty \\ \lim_{x \to c} g(x) = \pm\infty \end{array} \right) \quad \Rightarrow \quad \lim_{x \to c}\frac{f(x)}{g(x)} = \lim_{x \to c}\frac{f'(x)}{g'(x)}$$

$$\left. \begin{array}{l} \lim_{x \to c} f(x) = +\infty \\ \lim_{x \to c} g(x) = -\infty \end{array} \right) \quad \Rightarrow \quad \lim_{x \to c}\left[f(x)+g(x)\right] = \lim_{x \to c}\frac{\dfrac{1}{g(x)}+\dfrac{1}{f(x)}}{\dfrac{1}{f(x)g(x)}}$$

$$\left. \begin{array}{l} \lim_{x \to c} f(x) = 0 \\ \lim_{x \to c} g(x) = \pm\infty \end{array} \right) \quad \Rightarrow \quad \lim_{x \to c} f(x)\cdot g(x) = \lim_{x \to c}\frac{f(x)}{\dfrac{1}{g(x)}} \quad , \quad \lim_{x \to c} f(x)\cdot g(x) = \lim_{x \to c}\frac{g(x)}{\dfrac{1}{f(x)}}$$

$$\left. \begin{array}{l} \lim_{x \to c} f(x) = 0 \\ \lim_{x \to c} g(x) = 0 \end{array} \right) \quad \Rightarrow \quad \lim_{x \to c} f(x)^{g(x)} = \lim_{x \to c} e^{g(x)\ln f(x)}$$

$$\left. \begin{array}{l} \lim_{x \to c} f(x) = +\infty \\ \lim_{x \to c} g(x) = 0 \end{array} \right) \quad \Rightarrow \quad \lim_{x \to c} f(x)^{g(x)} = \lim_{x \to c} e^{g(x)\ln f(x)}$$

$$\left. \begin{array}{l} \lim_{x \to c} f(x) = 1 \\ \lim_{x \to c} g(x) = \pm\infty \end{array} \right) \quad \Rightarrow \quad \lim_{x \to c} f(x)^{g(x)} = \lim_{x \to c} e^{g(x)\ln f(x)}$$

www.matematicus.com

Function	Derivative
$y = k$	$y' = 0$
$y = x^{\alpha}$	$y' = \alpha\, x^{\alpha-1}$
$y = \sqrt[n]{x}$	$y' = \dfrac{1}{n\sqrt[n]{x^{n-1}}}$
$y = a^{x}$	$y' = a^{x}\ln a$
$y = e^{x}$	$y' = e^{x}$
$y = \log_a x$	$y' = \dfrac{1}{x}\log_a e$
$y = \ln x$	$y' = \dfrac{1}{x}$
$y = \sin x$	$y' = \cos x$
$y = \cos x$	$y' = -\sin x$
$y = \tan x$	$y' = 1 + \tan^2 x = \dfrac{1}{\cos^2 x}$
$y = \cot x$	$y' = -\left(1 + \cot^2 x\right) = -\dfrac{1}{\sin^2 x}$
$y = \arcsin x$	$y' = \dfrac{1}{\sqrt{1-x^2}}$
$y = \arccos x$	$y' = \dfrac{-1}{\sqrt{1-x^2}}$
$y = \arctan x$	$y' = \dfrac{1}{1+x^2}$
$y = \operatorname{arc}\cot x$	$y' = -\dfrac{1}{1+x^2}$